EFFECTIVE GRAPHIC COMMUNICATION

Norbert Lloyd Enrick, Ph. D.

Professor of Management
Kent State University

AUERBACH®
publishers

princeton
philadelphia
new york
london

Library of Congress Catalog Card Number: 70-171087
International Standard Book Number: 0-87769-097-9 Cloth
0-87769-078-2 Paper
First Printing

Printed in the United States of America

Contents

Preface		v
Acknowledgment		vii
1	Introduction	1
2	The Mechanics of Charting	9
3	Fundamental Graphic Forms	25
4	Sophisticated Charts	47
5	Flow Charts	67
6	Decision Charts	86
7	Charting Grids	97
8	Tabular Presentation	117
9	Decision-Structure Tables	137
10	Projection of Graphs and Tables	146
11	Preparation of Source Credits	152
12	Forms Design	161
13	Epilogue: Data Analysis and Communication—Today and Tomorrow	166
Glossary		173
Bibliography		180
Index		183

To the man who seeks simplicity and clarity
in his presentations, while yet coming
to grips with meaning and content

Preface

Students and workers in the technical, professional, business, and scientific fields have numerous opportunities to communicate their findings and ideas, while at the same time, people demand factual support of analyses and conclusions before results can be accepted or decisions made. The supplier of these analyses—the communicator—is often called on to buttress his message with informative and persuasive charts, graphs, and tables.

The purpose here is to aid the communicator in this endeavor by providing guidelines for good chart design and by being a source of ideas for effective and attractive visual and tabular forms. This book emphasizes:

Making the message convincing by stressing pertinent facts and factors and by minimizing the chance of tangled logic or confused presentation.

Competing effectively with the large amount of visual and reading matter with which the intended audience has to cope, by making charts, graphs, and tables interesting and attractive.

Holding the reader's attention with concise and precise forms, while at the same time presenting comprehensive and comprehensible content.

Presenting results, making the point, and conveying the message in a clear-cut forceful manner, as the overall outcome of good form, resourceful use of media, and meaningful information.

In working with this book, the user will soon discover that there is much more than the effective portrayal of data to be gained. The communicator, working thoughtfully with his media, will find that in the process, his ideas and understanding of the subject matter he is presenting have been sharpened. New insights into the working relations and other aspects of the data will be gained. Heightened awareness and knowledge are likely to be by-products of overriding value. Both communicator and audience will benefit from the improvements generated.

The task at hand is to study the material in this book, adopt the recommendations that apply to the reader and his needs, and make use of them in daily charting, graphing, and tabular design practice. In time, as experience is gained, the principles and techniques of good design will become an integral part of work habits and routines, with the resultant benefits to the communicator and his audience.

Kent, Ohio   Norbert Lloyd Enrick

Acknowledgment

Materials developed in this book are the result of over a decade of study of various forms of presentation. In this period, the author has had ample opportunity for practice. He has prepared over a hundred reports, research summaries, consulting surveys, and special analyses for management, and written a number of books. Inevitably, the author's thinking has been aided and shaped by the reactions, ideas, and suggestions received from his associates, colleagues, and clients during this work.

This book is a revised and greatly enlarged version of an earlier brochure, "Meaningful Charts and Graphs," prepared by the Center for Business and Economic Research in the College of Business Administration, Kent State University. Major acknowledgment is due to Dr. James E. Young, Director of the Center, for encouragement in this endeavor, as well as for reviewing, revising, and editing the earlier version. Finally, the Center also assisted in the preparation of a number of charts used in the present volume. Mrs. Mary E. Bacon, administrative assistant in the Center, helped considerably in the preparation, editing, and proofreading of several of the chapters. I am also indebted to Drs. Harriet Danielson, Charles Soltis, and Joseph Schwitter, professors in the College of Business Administration at Kent State University, all members of the editorial committee of the Center for Business and Economic Research who read the draft of the earlier brochure and made helpful comments and suggestions. Center funds also supported the research work of Mrs. D. L. Lewis as regards common usage of forms for source credits, such as those applicable to materials in tables and graphs.

Numerous publishers and authors granted permission for reproduction of graphic materials from their copyrighted works, and specific acknowledgments that include year of publication and copyright are given wherever such reused items appear.

1. Introduction

Inundated as we are by an ever-increasing outflow of data and information in almost every field of endeavor, we prize the man who can convey relevant facts, concepts, ideas, and relationships in comprehensive and readily understandable form. Charts and graphs have long been recognized as an excellent means of communication, because of their potential of concise, visual presentation of essential data and relationships. Good charts not only highlight the important aspects of a report or other document, they also stimulate the reader to delve into the detail contained in the accompanying text and tabular material.

THE PURPOSE OF CHARTING

Good charts convey information in concise and readily comprehensible form. Meaningful charts are those that reflect a real-world situation or a conceptual relationship in broad though faithful outline that highlights the essentials of the situation. While admittedly there are instances where charts are concocted deliberately to deceive or confuse an issue, the predominant purpose for which charts are prepared is otherwise: the author desires to enliven his material, highlight his findings, and ease the reader's task, all to the end of getting his message across and gaining acceptance for his ideas. A detailed look at the value of well-designed charts is given in Table 1-1.

NEED FOR CHARTING SKILLS

Despite the honesty and integrity of most authors, many times they will

TABLE 1-1

Value of Well-Designed and Executed Charts

General Category of Value	Underlying Basis of Value
Interest	The immediate appeal of a good chart invites reader attention, thereby creating interest in the information or message conveyed.
Portrayal of relationships	Visual comparison and contrast of pertinent data permits relationships to be more clearly grasped and more easily remembered.
Time saving	When large masses of data can be visualized at a glance, their essential meaning can be understood quickly. The underlying data may nevertheless be presented separately for further study and analysis.
Space saving	Graphic presentation of data, information flows, sequences and interplay of procedures and relations among concepts and other materials will generally require less time than plain text.
Synoptic overview	Comprehensive presentation of material in graphic form permits the reader to gain a quick total grasp of the essential information content. In turn, this promotes a fuller and more balanced view of the material than can be derived otherwise.
Unearthing hidden factors	Often, previously unsuspected relationships and other factors of importance are brought to the fore as a result of visual study.
Analytical thinking	Thought processes, analysis, ideation, and creativity can be enhanced by the graphic approach. This works two ways. The author, in designing and redesigning his chart, will be forced to think and rethink, thus stimulating his analytical, creative, and ideation-oriented faculties toward better, more complete and thought-provoking material. The reader, in viewing the final product, will often find in it food for further thought and analysis and possibly a source of new ideas.

2

fail completely in their efforts to convey their thoughts effectively. To some extent this failure is caused by inattention to certain mechanical prerequisites for graphic communication. But in a larger measure, many authors seem to have a lack of skill in designing charts. Yet just as writing and speaking abilities are the result of practice and experience, so charting requires a similar learning process to reach maturity. It is true that many simple types of graph require little knowledge or preparation, but for complex relationships, concepts, or thought processes a high degree of graphic skill is required. Where such abilities are not present, one of two results may occur. First, the author may forego the use of a chart because "the idea is too complex to chart," without considering that if he cannot graph the relationships among a set of concepts he is unlikely to convey much meaning by textual material alone. Second, the author may come up with a chart that fails to present his ideas clearly, thereby detracting from, rather than adding to, his overall effect. On the other hand, the author who has the proper charting skills will recognize where and how charts can and should be used effectively and will be successful in coming up with the right kinds and designs of visual material.

THE PROCESS OF CHART DESIGN

Assuming that a writer does have the requisite graphing skills, how does he go about his job? In instances of simple data, he might merely review in his mind the basic graphic forms available and then decide on the particular type that best suits his purpose. There may be some trials with different scalings, distances among lines or bars, symbols and designations, but the final form can be developed relatively simply.

When more complex materials are involved, such as relations among a set of factors, concepts, or ideas, his task will be a difficult one. For example, the author may need to prepare a report of his study, research, or other findings. He will run over the chief points to be made and single out a number of topics for visual underscoring. These chosen items should, of course, represent the key points to be contained in the report. For each topic, he may run through these procedures:

1. Think about a particular point or topic.
2. Make a preliminary sketch.
3. If the sketch is satisfactory, it may be drawn up in final form.

4. On the other hand, if the usual occurrence prevails, the first sketch will be found lacking, and a redesign of the sketch will be needed.
5. Several redesigns may be required until a satisfactory final form is accomplished.

This process, depicted in Figure 1-1, underscores the fact that a really good graph is likely to have undergone a great deal of painful work and rework.

FEEDBACK MECHANISM IN CHART DESIGN

A crucial aspect of the flow chart just presented is its feedback mechanism. As each sketch is prepared, we ask: "Does the chart convey the thought properly?" A "no" answer calls for a recycling through redesign and renewed evaluation until a "yes" answer is obtained. Most processes, whether in chart design, writing, or other activities, have a feedback—or, as it is sometimes called, a "cybernetic"—mechanism which acts as a reviewing, revising, and self-correcting device.

In chart design, who answers the question posed above? Initially, it would be the author. But the process is not that simple. After the author has drawn up his "final" chart, based on a "yes" response to the question, he should sample prospective members of his intended audience. Showing them the graph, he should solicit an explanation of the thoughts evoked and information perceived. Unsatisfactory results from this sampling will call for overhaul of the design in light of the sampling responses. The degree of care, thought, and circumspection that go into this cybernetic process may well decide the final impact of the graph. The maxim "hard writing, easy reading" thus applies not only to text, but also to visual accompaniment.

COMPLETENESS OF CHARTS

To the extent possible, charts should be complete in the manner in which they express ideas. We often find graphs that are rather cryptic, with ill-defined scales and abstruse symbolism. The author's idea is that the reader should read the text and then, by switching back and forth from the textual pages to the diagram, eventually dig out the meaning of the graph.

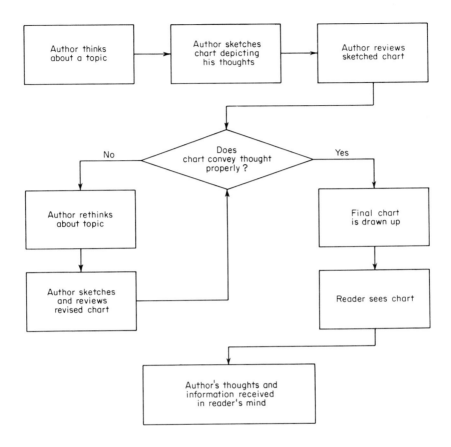

Figure 1-1: Cybernetic aspects of charting. Design of chart causes the author to review, rethink, and revise his own thoughts. The diagram that ultimately emerges is a vehicle for the transfer of thought and information from the author to his audience.

Such an attitude on the part of the chart designer will work with two groups of people, scholars who are used to such grief and students who are forced to do so because of a course requirement. For the most part, however, people in other walks of life will have neither the time nor inclination to put up with such chores. Picture an executive whose time is much in demand, who is harried by a multitude of problems, and who can only allocate a small amount of his effort to any one particular item. The high tension, multitudes of information flows, manifold interruptions, and weighty decision problems that surround him will hardly leave time for digging into scholarly endeavors.

It is thus a good rule that a chart should be self-explanatory, given the diagram itself, its title, subtitle, and brief explanatory text under the visual section. Despite some opinions to the contrary, this objective can always be accomplished, even if this sometimes means breaking up a single complex chart into two or three simpler ones. Regular textual pages will elucidate and supplement, but the effective graph will be largely self-explanatory. A certain level of technical knowledge on the part of the audience may of course be assumed in the design of the chart, provided this supposition is justified by the facts.

ACQUIRING CHART DESIGN SKILLS

The ability to design meaningful charts is the result of a learning process. Certain mechanics, ideas, guidelines, and illustrations may be provided. Thereafter, however, it is up to the individual to design his own charts, sample his prospective audience, and redesign his visuals until a satisfactory end result has been accomplished.

As in all facets of intellectual life, a continually self-reinforcing learning process occurs. With each successful chart, the designer's skills broaden, his creativity is enhanced, and his sense of accomplishment is strengthened. Moreover, the time required to prepare charts will shrink. Because humans are not fully aware of their own learning progress, the acquisition of new skills—even when it proceeds relatively rapidly—will often appear slow and "imperceptible." Nevertheless, the person who makes an honest effort to develop live, effective, and meaningful visual materials in his work will soon find that even complex ideas can be presented readily and quickly in graphic form.

Some people will fail to acquire good charting skills despite their honest desire to succeed. One predominant cause seems responsible for this problem: the hesitation and reluctance of many persons to put to paper something that they are not "sure" of as being "right." These habits, nurtured in their school days, are out of place in practical writing. To such people, the best advice that can be given is: "Put it down, by all means put it down on paper." After the work is in black and white, look at it for a while. You know it is not right yet, so erase some lines or blocks and put in better ones. After a while there will be so many erasures and lines that a new drawing is needed. So prepare it. Keep on going until you are satisfied. At first you may need a dozen drawings, but with experience you will discover that one or two will suffice.

In short, then, certain guidelines for effective chart preparation can and will be given. Useful ideas and illustrations are included. Thereafter, individual experience will serve to develop practice, skill, and professionalism. Thus most learning comes from doing.

DRAFTING VERSUS CHARTING SKILLS

The emphasis in this book is on the conception, design, and development of charts. The draftsman's skills in drawing up final diagrams, various means of inserting perfect lettering, and other such professional art work are not covered. Guidelines are given, however, for developing clear copy from which the draftsman or artist can work, in the case of books. For reports, where merely clean lines and typed-in lettering are considered adequate, the directions provided in the next chapter will be quite adequate. In no case should an author assume that an artist can take messy copy and prepare a good piece of final work from it. Since he is neither an expert in the technical subject matter of a report nor a mind reader, the artist cannot be expected to perceive the true intent of a messy graph.

CHART DESIGN AS A THOUGHT PROCESS

When complex ideas are reduced to visual form in a series of designs and redesigns, we do not just think about the diagram itself. Often in the process of examining and reexamining successive drafts, we realize that our original ideas need some revising, revamping or enlarging. The work of preparing the chart has thus resulted in our own rethinking and thus an enlargement and enhancement of the entire thought process surrounding the topics depicted graphically. This aspect is a most important by-product of the charting job. More than once, a writer may note a gap in thought, an omission of a vital factor, or an initial failure to recognize certain interrelationships only *after* he had reduced his ideas to lines and blocks on a diagram.

From this experience has come another maxim: each major idea in a report or other written material that involves a combination of factors, concepts, ideas, or interrelations should be reducible to graphic outline form. The diagram need not reveal all facets of the idea, but it should show the principal aspects and relations. The excuse that something is "too abstract" or "too complex" to be done graphically should be dismisseed as unacceptable. Force yourself and at least try to make a diagram. If you are alert, you will discover that your initial inability to prepare a meaningful graph is the result of gaps in your original thought processes. With persistence you will bridge the gaps, fill in missing links and eventually come up with the right diagrammatic representation of your ideas.

SUMMARY

Charts serve an author in livening up, underscoring, and effectively conveying the principal thoughts in his papers, reports, or other written or orally presented material. He thus aims at getting his message across in the strongest and most effective manner possible. Charts that are true to the author's thoughts and intent and are seen in proper perspective by the audience are thus meaningful charts, useful to both writer and reader. Skill in preparing good charts comes primarily from practice, although some helpful guidelines, ideas, and illustrative examples can be provided. An important by-product of charting is that it stimulates a person to re-examine his own thoughts and ideas, to make revisions and enlargements as necessary, and to consolidate and enhance the final product. Charting is thus as much an exercise in thought as the original study and the writing of the text, for which the visual material serves as a vital supplement.

A note of explanation: While this introduction has emphasized charting, the principles also apply in a parallel though generally less dramatic way to tabular design. Charting is the principal topic of discussion of the initial and middle chapters of this book, while specific material concentrating on tables appears in the later chapters. By this sequencing the author does not mean to imply that tables are less significant than charts as means of concise, comprehensive presentation. Rather, once certain principles are understood in the context of charting, their further application to tabular forms can be readily conveyed in less space.

2. The Mechanics of Charting

While much ideation, creativity, and ingenuity may have to go into the design of a chart, this work is unlikely to be fully satisfactory unless the humbler aspects of the mechanics of preparation are known, understood, and applied properly. Moreover, this detail of execution of graphic material should come into play at the early sketching stages.

PROBLEMS ARISING FROM POOR MECHANICS

Disregarding the mechanics involved in the charting process will not only handicap the author in his design, but will surely cause lost time and motion throughout the successive stages from original ideation to ultimate publication.

For example, a poorly prepared sketch may hide from the author's recognition certain gaps, omissions, or other deficiencies of content. Essential relationships among factors may remain obscured. A clear draft with proper mechanics might have revealed these problems and thus permitted corrective action. As another example, assume a basically complete design, sketched with improper mechanics regarding scaling, symbol identification, and readable lettering. When the final draft, in poor condition, is shown to a sample of the eventual audience, it is likely to be misread and misunderstood. Finally, a poor design may give all sorts of headaches to the draftsman or artist who has to try to interpret or "guess" what the author had in mind. Misinterpretations are bound to occur and to carry forward into the final drawing.

More than just time and effort may be lost. Publication deadlines are often quite close. When poorly conceived drafts result in a large number of erroneous final pieces of art work, there may be neither the time nor the requisite budgetary allowance to do the work over.

The types of problems noted are quite needless, since the habit of good mechanics can be easily acquired. In fact, it takes no longer to do mechanically correct sketches than to do them "any old way." What is needed most is a resolve on the part of the author to pay to these humble aspects the attention they deserve.

ESSENTIAL CHARTING MECHANICS

The mechanics of preparing good sketches that are "technically correct" from the start, even though the content may require several revisions, involves a number of factors. These are listed in Table 2-1, which serves as an outline of the principal considerations involved. It is advisable to review the list carefully and to refer to it from time to time as you design and redesign your own charts. Only with persistent effort can these details become an integral and routine habit of work, thus permitting a reasonable degree of precision in chart preparation right from the first sketch and leading to an overall improvement in design.

In the following we will examine several categories of this list in further detail, keeping in mind the author's need for effective, practical, and time-saving procedures.

MARGIN REQUIREMENTS

When preparing sketches, the usual sheet of paper of $8\frac{1}{2}$-inch width and 11-inch length is not only the handiest, but often the most suitable to use. Not all of this area can be consumed by the diagram itself, since we must allow room for the following:

1. Margin, usually on the left-hand side, for binding.
2. Margin, at top or bottom or both, to insert page number, chapter title, and other general information called for by the style of the report or other document.
3. Additional margins, at other edges of the paper, to make the diagram aesthetically appealing.

The margins shown in Figure 2-1 are thus generally minimal, not maximum, allowances. It should be realized further, however, that the

TABLE 2-1

Charting Details to Watch in the Design Stage

Category	Requirement
Margins	Allow margins to permit binding. Aesthetic appeal will be enhanced if the borders of the graph are at least one inch from the edges of the paper.
Text	Hold explanatory or identifying text appearing on the chart to a bare minimum. Edit and reedit text until it has been trimmed to fit the available space. Remember that supplementary explanations can be given below the graph, following the title and subtitle.
Lettering	Avoid oversized lettering that dominates the graph. Keep lettering large enough, however, to be easily readable (after allowing for reduction in size that may occur in the published version).
Scales	Place scales to allow room for titling them and indicating the units of measurement. Dimensioning and intervals between markers (or "ticks") should be wide enough to permit entry of scale values.
Titling	As noted, each scale should be identified by a title. Next, the chart itself receives a title (usually, but not necessarily, placed below the chart). Keep the title short to permit ready reference. For further description, use a subtitle, which follows the main title. Additional explanatory text may follow the subtitle. Edit and reedit all of the textual material for brevity and clarity.
Symbols	All symbols used in the chart should be identified as to their meaning. The identifying key may appear on a corner of the chart or below it.
Abbreviations	Treat abbreviations in a manner similar to symbols.
Instructions	Instructions to draftsman should be distinguished from other aspects of a graph by use of (1) different colored writing, or (2) placing the instruction material in a penciled cloud-like encirclement. Otherwise, the artist will not know what material is or is not part of the final graph. Clear and legible text, lines, and scales will be most helpful, also.
Review	Review each completed draft. Does it fully and concisely reflect the ideas, message, and intent you had in mind? Revise where necessary. Also consider breaking up one complex chart into two or more simpler graphs. Get other people's reactions before the final chart is drawn.

11

borders depicted do not yet give the space on which the diagram will appear. Further allowances must be made for the following:

1. Insertion of scale values and scale titles for vertical and horizontal scales, including an indication of the units of measurement.
2. Title, subtitle, and explantory text pertaining to the graph as a whole.
3. Identification of symbols and abbreviations, either in a separate key or as part of the explanatory text.
4. Source references, where required, for the diagram itself or for the data plotted.

An example of a diagram placed within the confines just established is given in Figure 2-2. The title is given as "Monthly sales of widgets for year just completed"; there is no subtitle, and the further explanatory text reads: "Data show cumulative number of units sold and the percent of year-end total represented by each month's sales volume." The title is important particularly for subsequent reference purposes, such as when a list of diagrams is provided. Although there is a figure number associated with this graph, such is usually not enough. Admittedly, in many writings the author will omit both title or explanatory text. In such instances, the reader is forced to scurry though the text to glean the meaning of each chart.

LETTERING

Letters that are too large will tower over and dominate a chart, thus detracting from its main purpose. A good rule to follow is that lettering should be readable, distinct, and clear, but not overly bold. Since most drawings are usually reduced photographically for final publication, consideration must be given not to the letter-size in ink form but to the ultimate readability in reduced dimensions.

SCALES

On most graphs at least two scales will be needed, so that the horizontal axis and at least one vertical axis are graduated. At regular, proportional, or otherwise determined intervals, scale markers, or "ticks," will be inserted. Enough of these ticks should be identified by values for the user of the graph to be able to determine (at least approximately) the scalar values associated with any point on the graph. For example, assume a scale of

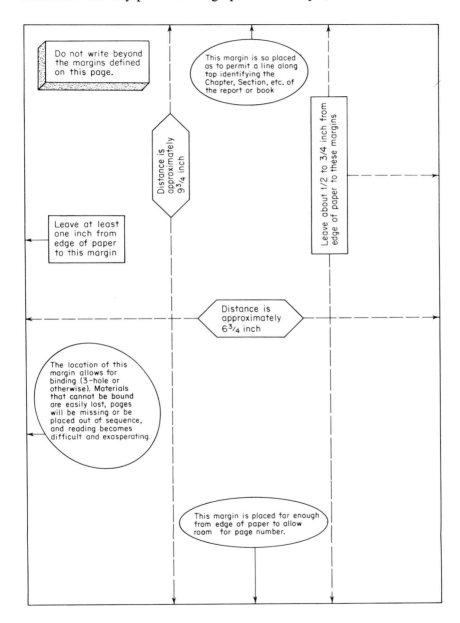

Figure 2-1: Margins for tables and graphs.

breaking strength values in pounds, ranging from 50 to 90. There is a tick mark every 5 pounds along the scale, at 50, 55, 60, etc., until 85 and 90. If every tenth pound is identified by value, such as 50, 60, 70, 80, and 90, the user can readily note that the intervening values will be 55, 65, 75, and

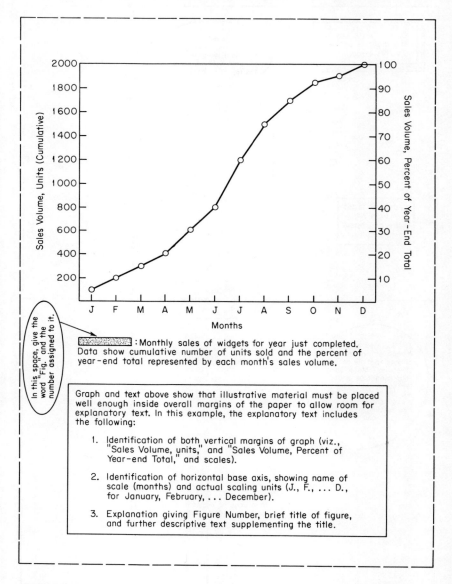

In this space, give the word "Fig." and the number assigned to it.

⬛ :Monthly sales of widgets for year just completed. Data show cumulative number of units sold and the percent of year-end total represented by each month's sales volume.

Graph and text above show that illustrative material must be placed well enough inside overall margins of the paper to allow room for explanatory text. In this example, the explanatory text includes the following:

1. Identification of both vertical margins of graph (viz., "Sales Volume, units," and "Sales Volume, Percent of Year-end Total," and scales).

2. Identification of horizontal base axis, showing name of scale (months) and actual scaling units (J., F., ... D., for January, February, ... December).

3. Explanation giving Figure Number, brief title of figure, and further descriptive text supplementing the title.

Figure 2-2: Placing material within overall margins for tables and graphs.

85. Yet these latter values may have been purposely omitted, since an excessively dense clustering of numbers will have a confusing effect. Moreover, by estimate, such intermediate values as 56, 57, 58, and 59 may be approximated reasonably well. The purpose of a graph is not to give precise detail, such as would be provided in a table, but rather to show general trends and relationships. Any attempt to provide excessive detail in the scaling will thus defeat the primary purpose of the graph.

TITLING

A scale always has a name; it should be titled. Thus a scale showing breaking strengths would be titled "Breaking Strengths," and the additional words "in Pounds" or "in Pounds per Square Inch" would indicate the units of measurement of the data involved. Titles should be as complete as possible, but again, much thought must be given to accomplishing this objective within limited space allotments. It is thus not unusual for a scalar title to be revised repeatedly until this goal is attained.

In addition to the scales, the diagram as a whole will also need a title. The title should preferably be brief but adequate, to identify and distinguish the chart from others in the report, book, etc. Subtitles are optional and may be longer. They serve to further elucidate the intent of the main title. Finally, it is often desirable to provide a few lines of additional explanatory material with the title. The aim is to make the chart as self-explanatory as possible.

USE OF FULL PAGES

The margins for placement of a chart, as previously provided, involve use of a full page. The question is often raised whether a full $8\frac{1}{2}$ x 11-inch sheet is too much space to allocate to a diagram. The answer is that usually a page is just right. When a chart is very small, then it is likely to be either unreadable or incomplete or not large enough to present the author's idea fully. Also, consider that the diagram itself is meaningless without proper scaling, titling, identifying text, and other such devices. Add to this the space required for binding, figure and page numbers, source references, and other detail, and the larger part of a full page will almost always be needed.

Even when a diagram is clearly so small as not to warrant a full page in manuscript form, it is still desirable to show it on a single page. A manuscript will usually go through at least one rewrite. If a diagram is spaced within the manuscript text material, then each rewrite means that not only the regular text on that page but also the diagram must be redone. Yet only the textual part of the manuscript and not the diagram may need change. Thus, to avoid needless work, each diagram should always appear on a separate sheet until the final version of the report is prepared.

One rule may be helpful: if your sketch does not fill a page, review it carefully. Have you left out some important aspects or facets that really should be given with the chart? Are your scales too condensed? Is there enough space to show the various lines of the diagram clearly and distinctly from each other? Should more explanatory material appear under the visual portion of the chart? The answer to some of these questions may well be yes, and the resultant final diagram will then be more completely descriptive of the idea, concepts, suggestions, relationships, or other results you wish to convey.

SHOULD DIAGRAMS BE COMPLETE?

It is sometimes argued, "if you make it too easy for the reader, by showing all your ideas graphically and in self-contained, complete form, he will not bother to read the report itself." If this were true, then hurray for the graphs! They are a testimonial to the quality of thought that has gone into their design. But the body of text will not be redundant. There will surely be many readers who, stimulated by a review of the graphs and charts, will feel the desire and interest to study the entire material in further detail.

It is true, of course, that well thought out visual materials give an effective and comprehensive overview of a report and its principal ideas and results. Self-explanatory diagrams thus give the reader a tremendous flexibility of choice in absorbing only as much of the detail as he feels he requires for his purposes.

GRIDS FOR CHART LAYOUT

In designing graphs it is wise to use paper with preprinted grids of various types (several standard forms of which are provided in Chapter 7). The vertical and horizontal guide lines of the grid provide a useful basis not only for general chart layouts but also for drawing lines and curves and plotting points.

An application of preprinted grid paper is shown in Figure 2-3. Note that the grids are used for layout and sketching only. In the final diagram, dense or heavy grids are distracting. Most published diagrams, therefore, show little or no background grid work. When the draftsman or artist is commissioned for the final design, he will usually be instructed merely to indicate the position of grid lines as tick marks on scales (see Figure 2-3) or to draw in only the principal lines of the grid as a fine, light background. When reports are prepared without the draftsman or artist, a practical approach to use is given by the procedures illustrated in Figure 2-4. This method involves a small amount of time and yields neat, professional-looking end products for reproduction on various types of office duplicating equipment.

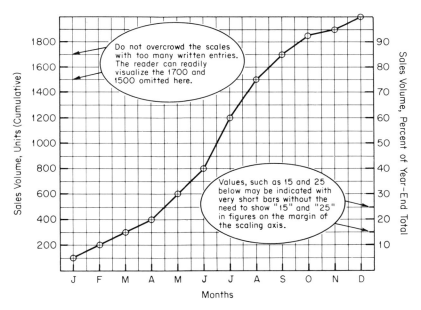

Figure 2-3: Scaling indication on graphs. Only the indications, not the background network of plotting lines, should appear in the final version.

OVER-SIZE DIAGRAMS

Occasionally it will be found that standard size paper is not adequate for certain graphic purposes. When this contingency arises, one should first review the data to ascertain that a larger size paper is really needed. Assuming that the answer is affirmative, it would then be an unwise and regrettable procedure to use just any other large size that may be available. A good approach is to use a double-sized page. For this purpose you should do one of two things: (1) purchase the right size paper in double-page format, or (2) paste two standard size pages together vertically. The second approach is accomplished by having one inch of the left-hand margin of the sheet to overlap the corresponding one-inch segment of the right-hand margin of the second sheet. The expanded paper thus obtained is pasted together at the one-inch overlap, thereby fitting well into the desired dimensional requirements. After three-hole punching and folding,

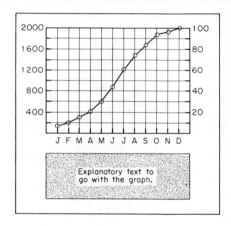

Explanatory text to go with the graph.

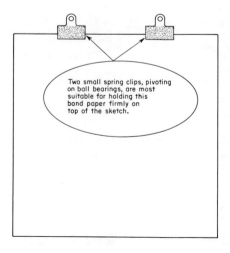

Two small spring clips, pivoting on ball bearings, are most suitable for holding this bond paper firmly on top of the sketch.

the new enlarged sheet fits into a report of regular $8\frac{1}{2}$ by 11-inch pages. This procedure is illustrated in Figure 2-5.

The use of off-size pages should be discouraged because they are likely to raise problems of proper folding and assembly into reports. Most office duplicating equipment cannot handle off-size paper, whether too large or too small. Unacceptable delays may occur while secretaries and clerks struggle with such nonstandard materials. Double size pages, obtained by means of the pasting procedure described above, are of course also difficult to deal with; but they are usually the least objectionable of the off-size, oversize problems that may arise. Of the many ways in which the chart designer can paint himself into a corner, the needless use of nonstandard diagrams is one in particular that should be avoided.

---

1. Sketch your diagram completely and as neatly as feasible on grid paper, with network or scaling that best suits your requirements.
2. The resultant diagram may, upon completion, look something like the example shown in reduced size on the left.
3. Be sure you have allowed enough room for all of the descriptive material—scale designations, title of figure, text to go with the graph, and other data belonging with the illustration.
4. Place completed sketch under a sheet of bond paper.
5. There are many types of good bond paper that will allow the sketched material to show through clearly.
6. Trace sketched material onto the bond paper. The paper will now contain your final copy, but without the underlying grid.
7. By using two types of pens (either draftsman's pens or pens with ordinary plastic tips), one light and one heavy, you can draw heavy and light lines as required for additional clarity of the graph.
8. While tracing, hold both pieces of paper firmly in place. Two small, thin spring clips, preferably pivoting on ball bearings, will guard against slippage.
9. Lettering of words and numerical values may be done by hand, but is more effective with a typewriter. The sketched-in lettering need not, of course, be as neat as the final version. Make sure that letter and figure sizes in the final diagram are neither so small as to be unreadable nor so large as to dominate the graph, thereby distracting attention from the main points of the diagram.
10. Make a final recheck to insure that there are no errors or omissions anywhere.

Figure 2-4: Preparing in-house reports without the aid of a draftsman. Attractive, neat, and grid-free material can be developed in a minimum of time by means of the technique described in this 10-step procedure.

USE OF BORDERS

Should diagrams be surrounded by a border, thereby setting them off clearly and distinctly from tabular and textual material in a report, article, or other publication? The answer to this question cannot be given in a clear-cut manner, and the choice is generally up to the designer.

In reviewing published material, it was noted that predominant usage

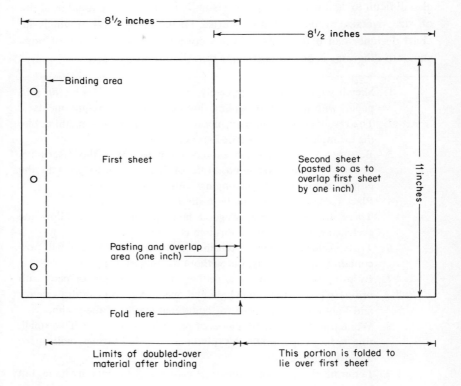

Figure 2-5: Dimensions for double-sized page. For many purposes, the diagram may be reduced to half size by photographic means (especially for books and brochures). For in-house duplication of reports, it may be feasible (though time-consuming) to prepare each page on a separate stencil and then do the pasting. Problems of oversized diagrams (excessive reduction in size or bulky page dimensions) make it desirable to limit them to instances of real necessity. Note that the scheme shown here permits binding the page within the confines of all other pages, thus minimizing handling problems.

is not to border off the graph or chart with an all-enveloping block. Accordingly, the procedure followed in this book has been to omit such framing. Nevertheless, there have been a number of instances when it was found desirable to use borders, particularly when it was felt that there was a real need to distinguish diagram from text. Figures 1-2 and 2-5 are two such applications. Both involve considerable diagram-related text, and the setting off of this material from regular text by means of an overall border was considered desirable. Objections may be raised that this approach is not consistent, because some diagrams now appear in separating frames and others do not. In reply to this objection, it may be pointed out that a consistent rule was indeed applied: separation is used for charts and later on for tabular material whenever such action is needed to insure clarity.

In practice, solid or dotted lines may be used for off-setting borders on charts and tables. In all instances, borders should be placed so as not to interfere with other matter such as the title of the diagram and the page number.

HUMAN RELATIONS PROBLEMS

Development of the proper attitudes in dealing with the mechanics of sketch and chart development often runs afoul of a persistent human problem. Fundamental to this is the author's frequent tendency to refuse to acknowledge any need to pay close attention to mechanical detail. After all, he argues, he is the thinker. His work is the hard work of analysis that culminates in the presentation of his findings and recommendations. That is the big job. How the report gets into finished and bound form is mundane and routine stuff. Anyone can do that job. He shouldn't have to worry about it.

No one wishes to take from this person his sense of pride and accomplishment. His contribution is indeed the prime, though not overriding, consideration. But the author must realize that failure to attend to the final details of providing clear copy, properly spaced and free from omissions, may cause his creative work to be for naught—for the simple reason that the final product will not be good enough to be easily read, convincing, and persuasive.

By "clear copy" we mean material that is understandable to the typist, artist, or draftsman, as well as the intended audience. Only when this requirement is recognized and accepted will the communicator's

TABLE 2-2

Check List for Chart Design

Item No.	Item	Check
1	Title	
1a	Preceded by figure number.	———
1b	Placed below diagram.	———
1c	Brief but adequate for identification.	———
1d	Subtitle explains further.	———
1c	Title and subtitle clear and concise.	———
2	Scales	
2a	Scaling shown by tick marks, not so detailed as to become crowded.	———
2b	Scale values given (but often it is preferable to give values for alternate or fewer tick marks only).	———
2c	Scales clearly labeled (both horizontal and vertical axes).	———
2d	Units of measurement (lb., in., degrees, etc.) indicated.	———
2e	Labeling text properly positioned in relation to axis.	———
3	Lines and Curves	
3a	Each line or curve labeled.	———
3b	When lines or curves cross, avoid confusion. Use different shadings, colors, or other such devices.	———
3c	Data points distinctly shown.	———
4	Footnotes	
4a	Symbols and abbreviations explained.	———
4b	Sources given.	———
4c	Qualifying statements and supplementary notes provided.	———
5	Spacing	
5a	Generally, one chart per page (except in printed forms).	———
5b	Standard size paper, except where larger size is absolutely unavoidable.	———
5c	Double-size paper where regular size is inadequate.	———
5d	Margin for binding and enough room on other margins to enter page number, etc., as may be required from a mechanical and appearance point of view.	———

TABLE 2-2 (cont.)

Item No.	Item	Check
6	General	
6a	Meaning and message clearly presented for the intended audience.	———
6b	Self-explanatory qualities. Diagram, labeling, title, and subtitle should be clear, concise, and sufficiently informative, so that the material stands by itself, without need to refer to other text.	———
6c	Uncrowded and uncluttered appearance. (Is it preferable to use two diagrams in place of one crowded chart? Can we simplify?)	———
6d	Enough information on chart. (Avoid empty-appearing graphs. Is additional information on hand to be meaningfully included on chart? Should two or more charts be combined into one?)	———
6e	Proper proportioning. On a chart, when some sections are jam-packed with material and other areas exhibit unused space, then the arrangement of information flow is inefficient. A well-proportioned chart makes relatively balanced use of all available space. Lettering, too, should be proportioned, based on relative importance of titles, scale values, explanatory text, and the like. Also, avoid oversize lettering that dominates and detracts from the diagram itself, while at the same time assuring writing that is large enough to be read without difficulty.	———
6f	Design and redesign. Often, a chart needs several designs, including rewording of textual matter until it is satisfactory.	———

Note: Special circumstances may make it necessary or desirable to supersede general charting rules. Be sure that reasons are valid. Also, do not be careless with the final draft, lest the artist misinterpret and thus misdraw the ultimate publications version.

charts help convey and sell his results, message, and ideas to his intended audience. It seems no exaggeration that a good idea may fail or be delayed in its dissemination and acceptance merely because these human relations factors were not dealt with resolutely and effectively.

CHECK LIST FOR CHART DESIGN

Although good design practices should become second nature to all those who must design charts and graphs, it will take some time until all essential principles are utilized regularly and properly. The beginner is likely to overlook some detail of minor or sometimes major significance in his work. The check list for chart design in Table 2-2 serves as a worksheet and reminder. The completed diagram should be compared against the points shown on the check list. This practice should be continued until such time as the chart designer feels he has full command of the essential principles involved.

It should be emphasized that the check list is not an inflexible list. There may be good reasons for preferring rules that differ in various aspects and yet have a consistency all their own, useful for a designer's style, methodology, and philosophy of charting. For such purposes, the chart designer may well wish to develop his own modified check list. There are a few points, however, that need to be kept in mind:

1. There should be good, well thought out reasons for deviations from general practice. The advantages of such deviations should be weighed against the drawbacks of nonstandard procedures that may not be readily followed or understood by the reader.
2. There should be consistency of application.

As has been pointed out, no one expects to continue to refer to the check list once the principles and mechanics comprising the list have become an integral part of his work pattern. There is, however, still a further use that can be made—training others. Whether this is in a regular classroom setting or in instructing one's secretary to prepare a graph, or in instructing staff people in a department to prepare better charts to go with their reports, we can in all instances use the check list to point out to such persons the types of errors that may be occurring in their design and preparation work. The check list is thus not only a self-teaching tool, but also an aid in training others to use effective, proper, and consistent practices.

3. Fundamental Graphic Forms

Predominantly, the types of graphs and charts encountered are relatively simple, and it may seem like a waste of the reader's time to examine them. There is value in such an endeavor, from the point of view of completeness and inclusion of those definitions that have received wide acceptance for the various forms of presentation.

LINE CHARTS

As the name suggests, a line chart contains a set of points connected by a line, as illustrated in section A of Figure 3-1. The basis, or "abscissa," of the chart indicates five manufacturing plants, while the vertical axis, or "ordinate," reveals labor costs, in $100,000 increments, incurred. For example, Plant A spent $200,000, while Plant B spent $300,000. This information has little value for interplant comparison without knowledge of the size of each plant, the product made, the age of machinery in place, and similar data. We shall, however, not concern ourselves with this type of problem at the moment. A more serious criticism of the chart is that the term "$200,000 labor cost" is insufficient by itself, since we might at least expect an indication of the time period involved, such as "$200,000 per year." Often, however, such information is purposely omitted because (a) the time period is abundantly clear from the context of the entire discussion accompanying the chart, and (b) extensive detail given with a scale may overcrowd it and cause reproduction problems. Admittedly, these are matters of opinion, but a good rule to follow is to place as much information as possible (without undue crowding) right with the graph itself.

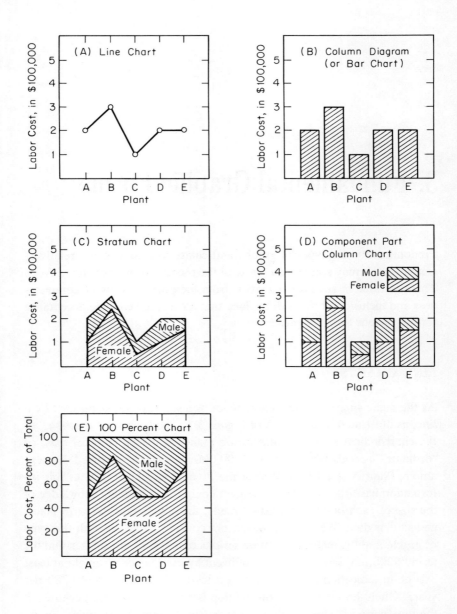

Figure 3-1: Line and bar charts. Note how the same basic information can be presented in various forms. The stratum chart is a detailed line chart, while the 100 percent chart restates the information given in the stratum chart. The forms here represent the most frequently used graphics.

26

COLUMN DIAGRAMS

When the points of a line chart are replaced by columnar bars, we have a "column diagram" or "bar chart," as shown in Figure 3-1, section B. Both types of presentation permit further segmentation, as shown in sections C and D, resulting in the stratum chart and the component part column chart. Finally, instead of a scale in actual values, we can make a conversion to percentage form. For example, if a plant spent $200,000 on labor costs, equally divided between female and male help, then on a one-hundred percent chart (as shown in section E), 50 percent of the height would be given to male and the remaining 50 percent to female help. Although the one-hundred percent chart is shown in stratum form, a parallel columnar version would be equally effective.

In our illustration the bars are shown in vertical, or "columnar," arrangement. It would have been just as feasible to extend the bars horizontally, beginning at various levels of the vertical axis. The result is a horizontal bar chart, a form widely used.

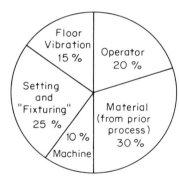

Data for Pie Chart			
a. Source of Variation	b. Index of Variability	c. Percent of Total Variation	d. Degrees = 360 × (c)%
Operator	4	20	72
Material	6	30	108
Machine	2	10	36
Setting	5	25	90
Vibration	3	15	54
Total	20	100	360

Figure 3-2: Pie chart, a popular method of presentation. This illustration shows the relative magnitude of various sources of variation in a manufactured product. For example, variability in the material fed from the prior process is the most important cause of variation (in dimensional or other characteristics) of the present product, followed by setting and "fixturing" effects. By emphasizing causal factors in this manner, they are highlighted for further study, thus giving priority considerations to relative magnitude. The method of obtaining the 360-degree pie chart is shown on the right. This detail is usually omitted, but if it is given, column *d* is left out.

PIE CHARTS

The one-hundred percent chart is a forerunner of another popular form of data visualization, the pie chart. Figure 3-2 illustrates such an application, together with the method of converting the original data first to percentages and then to degrees of a circle. The latter values, in turn, form the basis for the "slices" of the pie.

FREQUENCY POLYGONS

A special kind of line chart is the frequency polygon of section A, Figure 3-3. The crucial factor here is that the vertical axis represents not just any scale, but a scale of frequencies. The term "frequency" refers to "how often" something occurred. In our illustration of motor age at time of failure, we find that in a particular industrial plant a certain group of 39 motors of a given type had these failures.

1. Two failed at an age somewhere between 0.5 and 1.49 years. (It is assumed that no failures occurred before 0.5 years of motor age.)
2. Eight failed at an age somewhere between 1.5 and 2.49 years.
3. Twenty failed at an age somewhere between 2.5 and 3.49 years.
4. Six failed at an age somewhere between 3.5 and 4.49 years.
5. Three failed at an age somewhere between 4.5 and 5.5 years.

Observe that we have inferred considerable detail from the graph that is not directly apparent from the figures. The base scale lists the successive years 1 through 5. It is obvious, however, that motors do not fail exactly at the end of a precise time period; rather, they failed within a certain range or span. While we could have given that time span specifically, it would have needlessly cluttered the presentation without adding any essential detail. The graph does show that eight motors failed at (approximate) age 2, and since the scale is in yearly intervals, the reader can readily infer the span from 1.5 to 2.5—except that by convention we mentally replace 2.5 with the more precise though cumbersome 2.49 years in the event of a "tie."

In lieu of actual frequencies, percentages could have been given, such

as the percentage of motors failing at various ages. A relatively sophisticated usage also permits the development of probability scales. For example, if the frequency polygon represents a type of motor to be installed
at another plant, where they will operate under essentially similar conditions, then the percentages of failure from the past experience may now

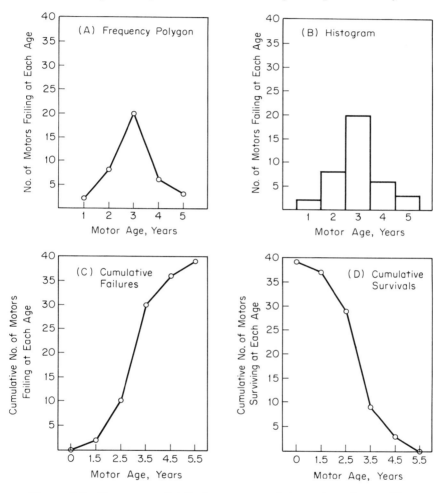

Figure 3-3: Frequency distributions expressed in various forms. The
term "frequency" arises from the "frequency" or "number of occur-
rences" indication given by the vertical scale. In place of actual numbers,
relative values such as percentage frequencies may be given. Note the
similarity of the frequency polygon to the line chart, while the histogram
is a type of column chart. Individual frequencies or cumulative occur-
rences may be used.

become "probabilities of failure, in percent" as an expectation of future experience. Life insurance annuity tables and graphs, for example, are based on this type of reasoning. Moreover, for an increasing number of industrial and business management problems, probability distributions are utilized.

FREQUENCY DISTRIBUTION

A frequency polygon is merely one form of expressing a "frequency distribution." The latter is a general term, applying to both tabular and graphic forms; it shows the frequency of occurrence of phenomena. This may represent the number of customers buying a certain size of soap powder, or the percentage of accounts receivable of various ranges of monetary size, or the probabilities of the volume of sales of a new product once it is marketed under certain pricing and promotion factors.

When plotting the motor age data, we utilized 5 groups or cells for the 39 frequency distribution values. This choice was not arbitrary. Rather, it followed a principle known as Sturges' rule,* from which the approximate guides below evolve:

Number of Values in Frequency Distribution	Approximate Number of Groups or Cells to Use
5 to 15	4 to 5
15 to 50	5 to 7
50 to 100	7 to 8
100 to 200	8 to 9
200 to 400	9 to 10
400 to 1000	10 to 12
1000 to 5000	12 to 14
5000 to 10,000	14 to 16

The more values (and thus data-points) there are to be presented, the larger is the recommended number of groups or cells.

---

* Sturges' rule is an equation from which the tabulation given here was derived. If N represents the number of values in a frequency distribution, then the number of cells C is found from

$$C = 1 + 3.3 \times \log N$$

The formula is intended only as an approximate guide.

HISTOGRAMS

An example of a histogram appears in section B of Figure 3-3. We note the similarity to a column chart, but here we are dealing with a representation of a frequency distribution.

CUMULATIVE FORMS

The frequency polygon may be represented in cumulative form, as shown in sections C and D of Figure 3-3. For example, 2 motors failed by the end of the "approximately one year" interval, which ends at 1.5 years. In the subsequent interval, 8 motors failed, thus giving a total of $2 + 8 = 10$ failures at the point when 2.5 years have been reached. The ascending line in section C illustrates this progression. A reversal of this reasoning leads to the curve for cumulative survivals, in section D. Thus 38 motors survived more than 1.5 years. In the next interval 8 motors failed, so that $38 - 8 = 30$ motors survived beyond 2.5 years. Next, 20 units broke down, leaving $30 - 20 = 10$ to remain functional.

TIME SERIES

When the base scale of a line chart represents a time scale, while the vertical scale is not a frequency, we are dealing with a time series, as in Figure 3-4. The particular illustration uses a double scale, thereby accommodating two distinct series.

There should, of course, be a good justification for placing two different sets of data on one graph. In the present instance, the preparer of the chart had noted that sales of the firm's plumbing equipment (one series) correlated relatively closely with new housing starts (other series). This relationship not only justified the joint graphing, but made the data more meaningful to the reader. Nevertheless, care must be taken in drawing conclusions regarding the relationship, and it would be highly desirable to supplement this graph with a correlation analysis, as will be discussed in the next chapter.

Often, several series of data will be found to be related. In such instances one should resist the temptation to place too many of them on one graph. A double graph, such as is shown in Figure 3-5, may be preferable. At what point should one avoid more lines on a graph? There is no hard and fast rule. Generally, if clutter, crowding, and confusion are encountered, the chart designer has probably exceeded the limit for number of lines or "information content" readily absorbable from a single graph.

The data contained in a time series are often analyzed statistically to discover underlying factors and components. Much of this work is based on the theory, particularly with regard to economic series, that a time sequence contains these components:

1.   An underlying long-term or "secular," trend.

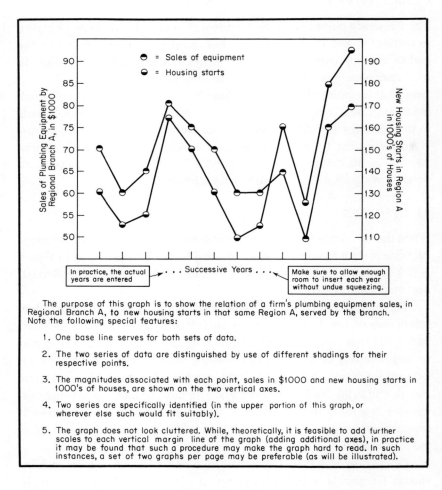

Figure 3-4: Time series. Two data series are given on one graph. Care must be taken to identify each set of data, so that the reader will not have difficulty in grasping the results.

2. Cyclical fluctuations such as the "business cycle" of prosperity and recession, moving around the trend.

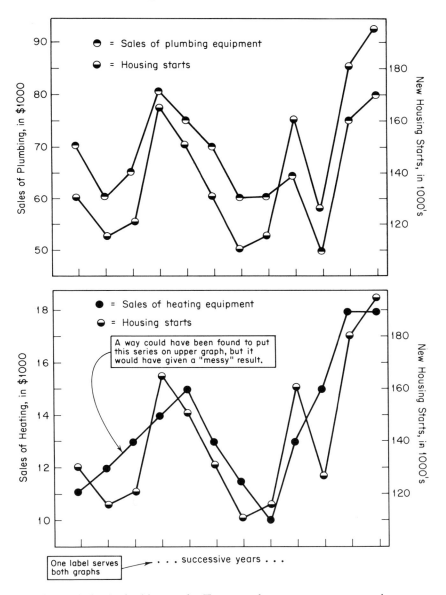

Figure 3-5: A double graph. Two graphs on one page permit easy comparison. Unless graphs are to be compared directly, however, it is best to use one page per graph in the manuscript stage.

3. Seasonal variation such as spring and fall peaks in economic, business, or sales activity.
4. Irregular or random fluctuations, usually small in magnitude, that cannot be readily explained.

Any observed point of such an economic or business series may then be viewed as the result of all of the four components above. In equation form:

$$\text{Any Point} = \text{Trend} + \text{Cycle} + \text{Seasonal} + \text{Random}$$

A more sophisticated formulation can also be developed, such that:

$$\text{Any Point} = (\text{Trend}) \times (\text{Cycle}) \times (\text{Seasonal}) \times (\text{Random})$$

A graphic view of these concepts is provided in Figure 3-6. While a detailed presentation of the methods of breaking down a time series into its components is beyond the scope of this book, it is helpful in viewing time series data to have a basic understanding of the four components generally recognized as underlying factors of such phenomena.

We may note in passing that Figure 3-6 contains four data series. Yet the presentation does not suffer from any cluttered effect, thus underscoring the point previously made, that no definitive rules apply as to the limit of number of lines per chart.

We will examine a number of further interesting aspects of time series graphs.

TIME-FREE BASE SCALES

Although most graphs contain a base scale in terms of time values, another sequence may be satisfactory. For example, we may be testing the hardness of a production lot of 100 castings, using a sample of 10 castings. In plotting the test results, our base scale will be Sample Unit No. 1, Sample Unit No. 2, and so forth, until Sample Unit No. 10. The chart will still be called a time series graph. Or we may record score points obtained from successively interviewing several dozen people as to their reaction to a new consumer product. Again, while the base scale will record successive respondents, the graph is referred to as a time series. As

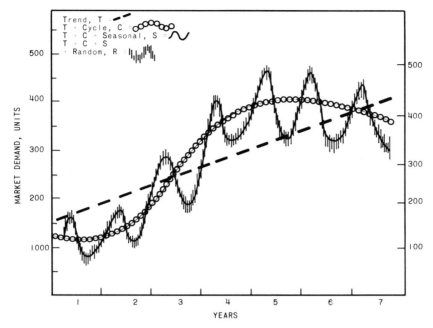

Figure 3-6: Components of a time series. From N. L. Enrick, *Market and Sales Forecasting,* Scranton, Pa.: © Chandler Publishing Co., a division of Intext, 1969.

a relatively extreme example, assume that you record sales volume in the 50 counties of a state. Dollar volume of sales will be scaled along the vertical axis of the graph, while county name will be indicated horizontally. Time per se is not involved in the graph, yet the particular arrangement of successive recording of data makes the appearance of the chart essentially a "time series," and such it is often called, even though a more specific term would be "line chart."

TIME-SERIES VERSUS FREQUENCY DISTRIBUTION

It should also be recognized that one does not simply plot a sequence of statistical data in time series form without considering the usefulness of this approach. Often a frequency distribution may be preferable. Take, for example, the motor age data of Figure 3-3. Originally the individual observations came in the sequence corresponding to time series data, that is, at various points in time individual motors failed. Plotted as a simple time series, this information would not have readily shown the more meaningful results, such as average age of motor at time of failure and range of variation around this average, that are immediately apparent from the frequency distribution.

CONVERTING ABSOLUTE TO RELATIVE VALUES

Finally, it will often be desirable to convert raw data to more meaningful indexes or other statistical values prior to plotting. Even where the raw data are portrayed graphically, further treatment may provide additional useful insights. Thus the plumbing sales and housing data of Figure 3-4 might be supplemented by another diagram showing only one vertical scale and thus only one line (in lieu of the present two) in terms of "Plumbing Equipment Sales in Dollars per 1000 New Housing Starts." The ratio so formulated then automatically shows the firm's sales performance in relation to the size of the market from year to year. In other words, we have converted absolute units to relative values in order to extract additional meaning from the data. Finally, where previously two time series lines were required, one is now called for. Now let us look at the double graph in Figure 3-5. It is apparent that both upper and lower sections could have been condensed by conversion to relative terms, as was just shown. By recombining the two resultant single-line time series, we would have obtained a single new graph with one time line and two vertical scales. To the right, the scale would read "Plumbing Equipment Sales in Dollars per 1000 New Housing Starts" and to the left we would have "Heating Sales in Dollars per 1000 New Housing Starts." Noting the repetition in the two quoted scale captions, we might simplify further by using an overall caption for the graph: "Sales Dollars per 1000 New Housing Starts," leaving "Plumbing Equipment" to the left-hand and "Heating" to the right-hand sides of the scaling indications bordering vertically.

UTILIZING STANDARD VALUES

Often it is not possible to obtain relative values from the actual data, because a multitude of variables affect a particular time series. Thus, as we noted in connection with the data of Figure 3-1, such factors as size of plant, product made, and machine age have a bearing on labor costs. Now in order to express labor costs in comparative and thus relative terms among the plants, one must first convert the factors of plant size, product made, and machine age to a composite value. Usually we develop a formula into which the variables are fed in order to obtain "standard labor cost" for a given period. Actual labor cost, when expressed as a percent of standard, then represents a relative value. Its usefulness for comparative analyses depends of course on the validity of the formulas from which the standards are obtained.

INDEX NUMBERS

A popular method for obtaining relative values in a time series is the "index number," that appears as the vertical scale. Let us assume, for example, that unemployment rate per 1000 population in a particular locality is as follows in successive years: 4.0, 3.8, 4.2, 4.8, 5.0, 4.4. Any year may now be taken as the "base year." Assume that the first year is used as a base and therefore receives an index of 100. The following year 3.8 is then:

$$\text{Index} = \frac{\text{Year Under Study}}{\text{Base Year}} \times 100$$
$$= (3.8/4.0) \times 100$$
$$= 95$$

Proceeding similarly for the other years, we obtain successively 105, 120, 125, and 110.

Although an index is a relative value, it is so usually with regard to some other value in the series. It may thus be deficient for other comparisons, such as among series or among values that are intrinsically different for other reasons—which may have caused us to express them as percentages or rates. Finally, when rates or percentages have been obtained they may be further changed into index numbers. Care must be taken that the reader, for whom the charts are intended, will be given information in adequate detail to understand, comprehend, and effectively use the results.

CONTROL CHARTS

When a set of control limits is given on a time series, the result becomes a control chart. The limits are usually calculated by means of statistical methods. We will not go into the detail of their derivation. Suffice it to say that the limits are set so that plotted points representing "normal, hour-to-hour or day-to-day fluctuations" expected for a particular time series will generally fall within the control limits. An out-of-control occurrence thus signifies that "something seems to have gone wrong." The exception principle of management is thus at work. Usually, staff, executives, managers, and supervisors will not worry about small fluctuations in successive data points. Only when a deviation beyond either an upper or lower control limit occurs will there be cause to check into the factors responsible and to seek corrective action.

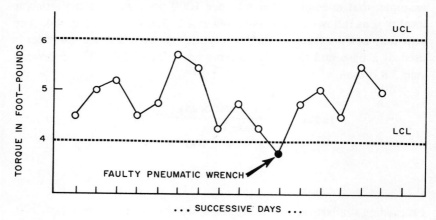

Figure 3-7: Quality control chart. The illustration refers to the assembly of fuel pump meter covers to bodies. Each plotted point is the average of torque tests, in foot-pounds, on 16 bolts. Note that both excessive tightening and loose tightening, as revealed by high and low torques, are undesirable. An out-of-control point calls for investigation and correction. Since leakage generally is not discovered until the pumps have been in service for some time, this quality control chart reduces the amount of field repair calls.

UCL = upper control limit; excessive assembly tightness means cover distortion and leakage

LCL = lower control limit; loose bolts permit leakage between cover and gasket

Let us examine several typical charts. Figure 3-7 is known as a control chart and applies to an assembly operation. We are concerned with unduly loose bolts (lower control limit, LCL) that would permit leakage and excessively tightened bolts that would mean cover and gasket distortion and again leakage (upper control limit, UCL). Figure 3-8 shows a

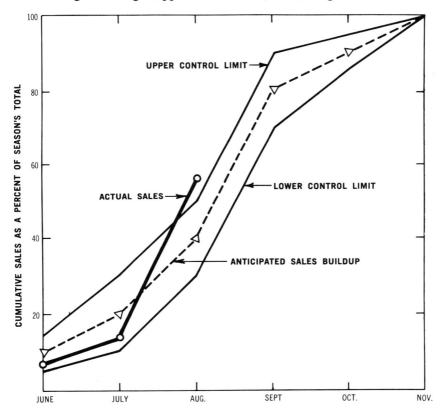

Figure 3-8: Control chart for cumulative data. Illustrative example refers to a seasonal anticipated sales buildup. When actual sales rise above control limits, action to provide for greater-than-expected market demand is called for. Conversely, actual sales below lower control limits caution management to consider cutbacks. A company may maintain dozens of charts like these, one for each product. Some products may call for cutbacks at the same time that others call for increases. Use of the charts affords valuable time in planning and providing for future needs as revealed by the market trends. Adapted with permission from N. L. Enrick, *Inventory Management,* Scranton, Pa.: © Chandler Publishing Co., a division of Intext, 1968.

(stock is falling too low). From N. L. Enrick, *Sales and Production Management Manual,* New York: John Wiley & Sons, 1964.

seasonal sales control diagram. When actual sales go outside the control limits, which are cumulatively plotted, one of two problems may exist. A point beyond the upper limit means that market demand tends to exceed earlier anticipations and quick action seems desirable to plan production of additional merchandise. Conversely, an actual sales trend falling below the lower limit calls for a cutback. A similar chart, for department store merchandising, indicating when it may be desirable to reorder stock or to mark down prices is given in Figure 3-9. Finally, Figure 3-10 shows administrative and financial applications of control charts as aids to executive evaluation of overall performance.

   We have been deliberately brief in discussing the manner in which management, cost accounting, and statistical analysts should combine their efforts to arrive at practical, economic control limits. This is a

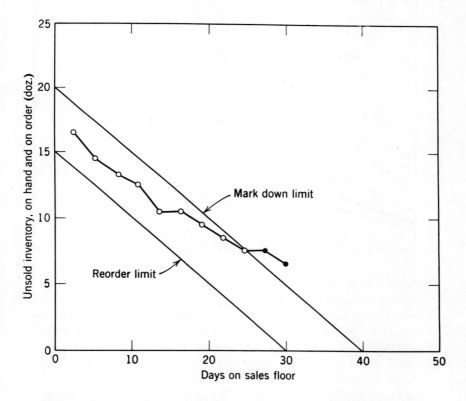

Figure 3-9: Control chart for department store merchandising. The upper and lower limits indicate when to consider "mark downs" of merchandise price (stock is not moving fast enough) or when to reorder

complex subject, generally presented under the topic of statistical control in modern quality control texts.* The entire concept of statistical control methodology in management has sprung from original applications

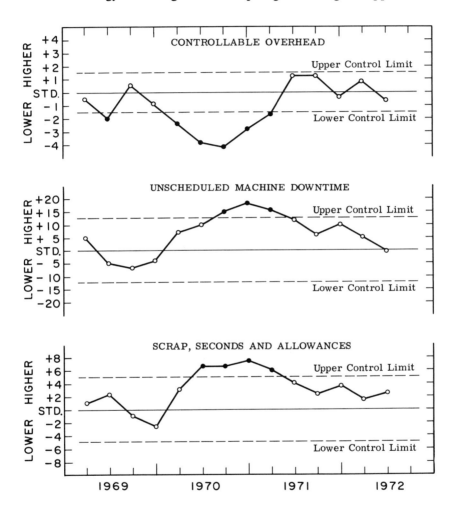

Figure 3-10: Financial and administrative control charts. Reviewing these, some readers may wonder whether an overhead cost can ever be "too low." But such is possible when, for example, inadequate maintenance and quality inspection are scheduled (top chart). Scrap, seconds, and allowances may then creep upward and out of control (bottom chart). The resulting inferior quality of product affects customer goodwill and thus selling expenses (middle chart).

* For example, N. L. Enrick, *Quality Control and Reliability,* 5th ed., New York: Industrial Press, 1966.

that were confined to the control of product quality and reliability, including waste reduction and productivity controls.

MODIFIED BASIC FORMS

A variety of modifications and adaptations are applicable to the fundamental types of graphic forms given in the preceding discussion. We shall now examine a number of predominant types.

STRATUM CHARTS

Portrayed in Figure 3-11, this is basically a series of curves on a line chart. Shading of the intervening segments results in distinct layers that contribute to a strata effect.

SILHOUETTE CHARTS

When fluctuating movements of a time series or other data occur around a straight line that serves as a reference base, it is often useful to show the deviations in shaded form. Figure 3-12 demonstrates this technique. The reference line may be a standard, a goal, an average based on long-term experience, or simply a zero level.

MODIFIED BAR CHARTS

The columns on a bar chart need not be flat. As shown in Figure 3-13, by placing a shading on sides of the columns, a three-dimensional effect is achieved. The specific techniques of achieving this type of presentation will become self-explanatory after a review of the next chapter, demonstrating the use of isometric drawings.

Another modification of bar charts is to use horizontal bars rather than vertical columns. The choice of a horizontal versus a vertical arrangement is usually optional, but in many instances one or the other design may fit better within given space allocations for the overall dimensions of the diagram.

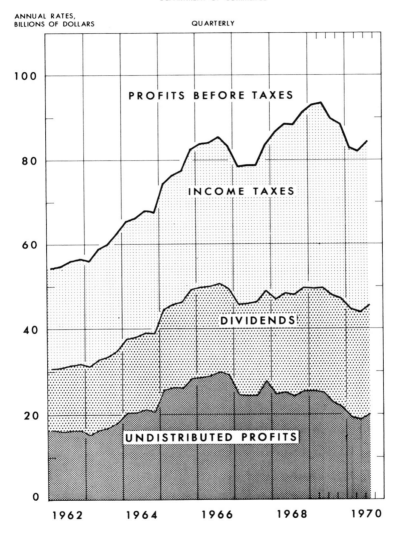

CORPORATE PROFITS, TAXES, AND DIVIDENDS

DEPARTMENT OF COMMERCE

ANNUAL RATES,
BILLIONS OF DOLLARS   QUARTERLY

100

PROFITS BEFORE TAXES

80

INCOME TAXES

60

DIVIDENDS

40

20

UNDISTRIBUTED PROFITS

0

1962   1964   1966   1968   1970

Figure 3-11: Stratum chart. Shaded segments show at a glance the composition of a total (profits before taxes, in this example) in terms of the component series (taxes, dividends, and undistributed amounts). From "Federal Reserve Monthly Chartbook, Financial and Business Statistics," Board of Governors, Federal Reserve System, Washington, D.C.

43

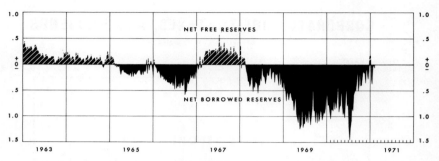

Figure 3-12: Silhouette chart. Deviations from a zero line are shaded. The zero line may represent a standard, a goal, or an actual zero. In this illustration zero represents a state in which a banking system has neither free reserves nor borrowed reserves. From "Federal Reserve Monthly Chartbook, Financial and Business Statistics," Board of Governors, Federal Reserve System, Washington, D.C.

MODIFIED PIE CHARTS

Pie charts may be presented in an interesting fashion by (1) showing them in flat rather than vertical form, and (2) utilizing actual physical items in lieu of a simple circle. Figure 3-14 brings an application of both of these ideas in showing first the individual sources and then the applications of income and expenditures per governmental dollar.

Aside from these types of modifications, it will be noted that all of the

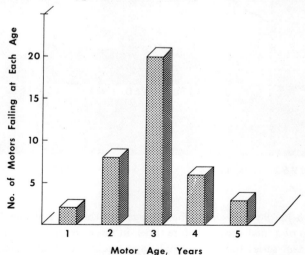

Figure 3-13: Bar chart with three-dimensional effect. Instead of a flat appearance, bars have been shaded so as to look like actual blocks. From National Science Foundation, "Reviews of Data on Science and Resources No. 12," January 1968, Washington, D.C.

Where It Comes from

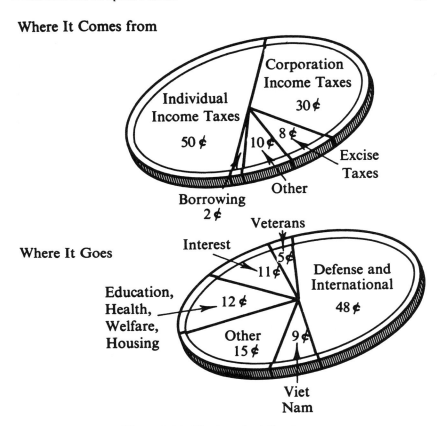

Figure 3-14: Pie chart in dollar form.

examples given in this section have a distinct styling that differs somewhat from that utilized in the execution of other charts in this chapter. The absence of universal standards and the right, as well as desirability, of the artist to show individuality and distinctiveness are thus underscored. One should make sure, however, that individuality stays within certain limits. Extremism may lead to a degree of abstractness that will create a real problem of chart interpretation on the part of the intended audience, which will detract from the message the communicator wishes to convey.

CHARACTERISTICS OF CURVES

Having examined the various types of fundamental graphic form, we can now consider the predominant characteristics in the movement or trends

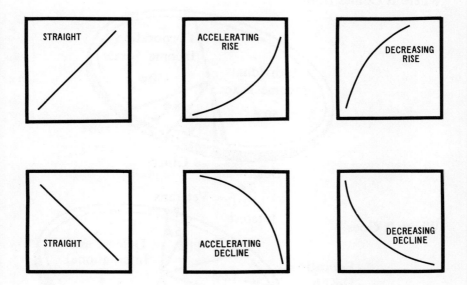

Figure 3-15: Characteristics of curves—the predominant ways in which curves may rise or fall.

of data, as depicted on charts. There may be level movement or erratic and inconsistent upward, downward, or sideways trends. Often, however, there are distinct patterns of increase or decrease, such as are portrayed symbolically in Figure 3-15:

1. Increases at
    (a) constant rate, giving a straight line.
    (b) accelerating rate, with a rapidly rising curve.
    (c) decreasing rate, with a rising curve that slows down as it moves up.
2. Decreases at
    (a) constant rate, giving again a straight line, but moving downward.
    (b) accelerating rate, with rapidly declining curve.
    (c) decreasing rate, with a curve whose decline lessens continually.

Real-life curves rarely exhibit these exact patterns from actual observations, but trends characteristic of the data can nevertheless often be recognized.

4. Sophisticated Charts

While the often heard admonition "keep it simple" is a meritorious one, it should perhaps be modified to read: "Keep it as simple as possible!" Often the nature of an idea, concept, or result is such that it is meaningful only in its totality, and "totality," in turn, may involve complexity. Charts and graphs can often help to clarify and simplify the comprehension of such material even though the diagrams themselves may by no means be simple ones.

MULTIFACTOR CHARTS

The multifactor chart is a presentation of many sets of results in comprehensive form, so as to facilitate multiple comparison. Figure 4-1 provides an illustration based on the results of an experiment designed to obtain the highest possible tensile strength, in pounds, for an asbestos cord. Several factors were considered in the investigation, each at different levels of intensity:

Factor studied	Levels of each factor	Number of levels
Fiber fineness	Coarse, medium, fine	3
Type of roll covering used in the processing operation	Cover A, Cover B, Cover C, Cover D	4
Twist inserted into the cord, in turns per inch	1.2, 1.4, 1.6, and 1.8	4

If we consider the three factors as "input variables," we have a total of four variables after considering "tensile strength in pounds" as the

47

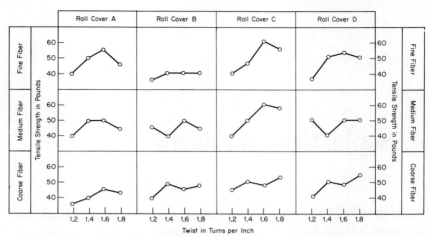

Figure 4-1: A multifactor chart. The results, in tensile strength of an asbestos cord, from a three-factor experimental study, are shown for unified, synoptic overview. From a review of (1) fiber fineness, (2) roll covering used during processing, and (3) twist imparted, the optimum combination(s) can be ascertained.

"output variable." Moreover, the three factors at the various levels give a total of $3 \times 4 \times 4 = 48$ "conditions," a result found by multiplying the "Number of levels" shown above.

The four variables and the 48 individual conditions to which they gave rise are all well accommodated in the illustrative diagram. It would have been a relatively simple matter to take the 12 individual "line-chart" segments of this comprehensive chart and represent them in separate graphs. But there would be many objections to such a procedure:

1. Showing 12 graphs instead of one would add burdensome bulk to the report, brochure, or book.

2. The reader would become annoyed by having to look at each line chart separately.

3. One would have to turn pages back and forth in order to discover, by tedious forward-and-backward comparisons, the optimal conditions.

4. We would lose the comprehensive, comparative overview among the various factors.

It is true that under either the single comprehensive chart or a set of individual graphs, we would soon uncover the optimal conditions. There are two of them, both leading to a strength of 60 pounds for the cord: (a) fine fiber, roll cover C, and 1.6 turns of twist, and (b) medium fiber, again roll cover C, and again 1.6 turns of twist. We may next also wish to consider other factors, such as production rates and raw materials costs involved in various factor-level conditions, before making a decision on what processing to use. Without the synoptic overview afforded by the comprehensive chart, it would be difficult to assess the strength-data in relation to cost or other considerations.

But there is still another reason for desiring comprehensiveness in charts. Often, from a thoughtful examination of the data trends and relationships revealed, we are able to decide on additional experiments, leading to further worthwhile results. Ideas for such investigations and the particular direction in which these should proceed are more likely to come from a comprehensive review of all factors than from a tedious, piecemeal, fragmented checking of results.

ENLARGEMENT OF MULTIFACTOR CHARTS

The illustration shown is not necessarily the ultimate in terms of factors that can be included in a multifactor chart that remains clutter-free. For example, suppose we had investigated fibers supplied by two different producers, Q and P. It would have been a simple matter to show the strengths from each producer's product, for the three finenesses, four roll coverings, and four twist insertions, merely by providing two lines on each of the 12 line-chart segments of the comprehensive graph. For each line, the points (small circles) would be shaded to distinguish producer Q from P.

Note that with the introduction of two producers we now have four factors. The entire system can be viewed thus:

Variables under study	*Levels of each factor*	*Number of levels*
Input variables or factors		
1. Fineness values	Coarse, medium, fine	3
2. Roll coverings	A ,B, C, D	4
3. Twist, turns	1.2, 1.4, 1.6, 1.8	4
4. Suppliers	Q and P	2
Output variable		
Strength in pounds		

Suppose, now, that strength is not the only consideration in making a decision on the ultimate processing setup to utilize in production. Then we can readily determine and graph, on separate multifactor charts, items such as:

1. Elongation characteristics of the product produced
2. Abrasive resistance
3. Productive capacity utilized
4. Labor cost consumed
5. Heat resistance
6. Flexures until failure

There may be other items as desired. Usually there are limitations to data-gathering, particularly when tedious experiments must be run and lengthy measurements taken, or where information has to be collated from numerous and splintered sources. People concerned with multifactor investigations will therefore try to confine their studies to the most important characteristics of a product. As an alternative, intensive investigations are made of factors that are considered likely to be most important, with briefer surveys for the remainder of items. Once all the information has been gathered, however, there is need for much thought about both tabular and visual presentation. Only this way can the results be readily comprehended for informed, optimum-seeking decision-making.

CALCULATION CHARTS

Sometimes it is desirable to combine calculations with the basic graphic presentation. Again, the objective is a ready overview without need to turn from graph to text or from graph to text and table. An illustration of this type of chart is given in Figure 4-2. The upper portion shows the frequency distribution of hardness test results, in terms of "degrees of hardness" obtained for a sample of 30 parts from a production lot.

For example, one part tested 6°, two parts tested 7°, three parts tested 8°, and five parts tested 9° on the applicable hardness scale of the test equipment. These results are recorded in lines *a* and *b*. We note that a total of 30 parts were tested. The simple calculations shown in line *c* lead to a total of 304, which is divided by 30 to give an average hardness of 10°, after some slight rounding. We are next interested in measures of

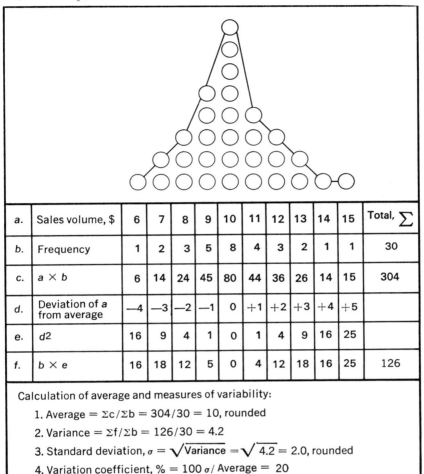

a.	Sales volume, $	6	7	8	9	10	11	12	13	14	15	Total, Σ
b.	Frequency	1	2	3	5	8	4	3	2	1	1	30
c.	a × b	6	14	24	45	80	44	36	26	14	15	304
d.	Deviation of a from average	−4	−3	−2	−1	0	+1	+2	+3	+4	+5	
e.	d2	16	9	4	1	0	1	4	9	16	25	
f.	b × e	16	18	12	5	0	4	12	18	16	25	126

Calculation of average and measures of variability:

1. Average $= \Sigma c / \Sigma b = 304/30 = 10$, rounded
2. Variance $= \Sigma f / \Sigma b = 126/30 = 4.2$
3. Standard deviation, $\sigma = \sqrt{\text{Variance}} = \sqrt{4.2} = 2.0$, rounded
4. Variation coefficient, $\% = 100\,\sigma/\text{Average} = 20$

Figure 4-2: Calculation chart. A frequency distribution is shown graphically. Below it are shown the basic values of the distribution (hardness, in degree, of metal parts in line *a*, followed by their frequency in line *b*). Next, the calculations are given for computation of the arithmetic mean or average and three measures of variability of the distribution, namely, variance, standard deviation, and variation coefficient. The advantage of this form is that the calculation procedures are tied directly to the visual presentation of the illustration to which they pertain. From N. L. Enrick, *Decision-Oriented Statistics*, Princeton, N.J.: AUERBACH Publishers Inc., 1970.

variability, and the three most predominant (and related) types are calculated in the remaining portion of the graph, lower section. First, in line *d* the deviation of each hardness value in line *a* is shown with reference to the average hardness of 10. Next, the values are squared and then further adjusted for frequency in lines e and f. The calculation of variance, standard deviation, and variation coefficient follows. Observe that all steps are made self-explanatory. This is accomplished by identifying all rows (in terms of lines a, b, c, etc.) and the totals column (large Greek sigma, Σ, denoting "sum of") and then demonstrating each calculation in terms of the numerical values identified by the row and column designations given. Another symbol, a small sigma, or σ, is also presented, but its purpose is merely to reflect customary statistical nomenclature.

We are not particularly interested in the detail of calculations, but are concerned with:

1. That methodology used represents a compact device for combining graphic visualization with demonstration of pertinent calculation procedures.
2. That data identifications (by rows and columns) permit self-explanatory presentation of the calculation steps in terms of formulas utilizing the identifications.

One further lesson emphasized by the chart is that there should never be any unexplained calculations on the chart if this is at all possible. Strict attention to the proper identification of all important sections (lines and columns) and unfailing use of formula representation of the calculations performed will insure this result. It should be realized, of course, that calculations may at times be too many and too complex to permit this approach. Nevertheless, the chart designer's first obligation to the reader is to try his best to give all pertinent detail, all essentials, and all calculations in one place—that is, *on the graph.*

STEP-SERIES GRAPHS

When an idea is too complex to be grasped from a single diagram, it is often desirable to have a series of step-by-step build-ups, such as in Figure 4-3. The problem involved in this example was one of sales-production coordination. How much of each of two products, A and B, should the sales department seek to sell per week in order to maximize overall profit, taking into consideration unit profits, production rates, and productive capacity? Because of the complexity of this task, the basic data and calculations must be shown separately (Table 4-1). Again, we are not so much interested in the calculation details as in the basic principles shown in following the table.

TABLE 4-1

Data Pertaining to the Step-Series Diagram in Figure 4-3
(Example represents a sales-production coordination problem,
seeking an optimum)

Values and calculation	product A	product B	Total
	Products Sales Department Considers Saleable		
a. Profit, $/gross	6	5	
b. Production rate, hr./gross			
polishing process	3	4	
plating process	2	1	
c. Productive capacity, hr./week			
polishing process			120
plating process			40
d. Output possible, in gross/week if *only* product A *or* B, but *not both,* is made ($= c/b$)			
polishing ($= 120/3$ and $120/4$)	40	30	
plating ($= 40/2$ and $40/1$)	20	40	
e. Bottleneck process ($=$ lowest gross/week for each product in d), which limits production of the other process	plating	polishing	
f. Output possible, in gross/week, considering the bottlenecks ($=$ output in d corresponding to bottleneck process)	20	30	
g. Profit, $/week, if only product A *or* B is made to the maximum possible ($= a \times f$)	120 (or)	150	150
h. Optimum combination of *both products,* from step 4 of graph, in gross/week	8	24	
i. Optimum profit, $/week ($= a \times h$)	48 (plus)	120	168

Notes: 1. Step d above yields the points, on the product A and B base lines of step 1 of series diagram, which are connected to show the processing capacities. The graph is next shown as the base of a cube (step 2), ready to add another dimension.
2. The profits obtainable from step g are shown by the columns leading to $120 and $150 of the graph in step 3 of the series diagram.
3. Maximum profit (step i) is represented in step 4 of the series diagram.

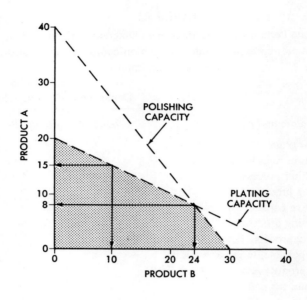

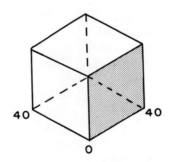

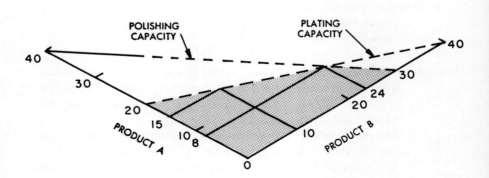

54

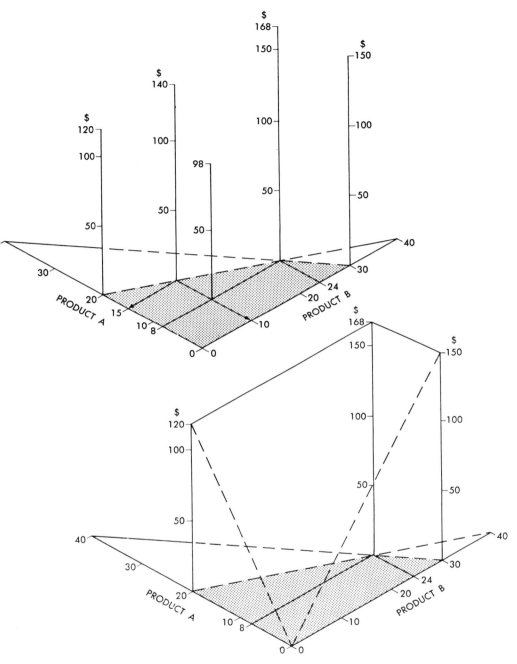

Figure 4-3: Step-series presentation. In this approach, a series of related graphs present an idea, beginning with a simple diagram and adding information until the total concept has been developed. This step-by-step approach facilitates reader comprehension of the ultimate, relatively complex graph. From N. L. Enrick, *Management Operations Research*, New York: Holt, Rinehart & Winston, 1965.

The basic principles:

1. The problem aspects are developed gradually, from the simplest form in step 1 to the most refined (step 4).
2. Data that probably cannot be readily visualized from the final step alone do become understandable from the step-by-step development.
3. A crucial change such as a shift from two dimensions (step 1) to three dimensions (step 3) is smoothed over by means of the intermediate illustration in step 2. The "cube" is designed to demonstrate the shift toward three dimensions, with the data from step 1 now shown as the base segment of a cube. The third dimension, profit in dollars, is then brought in readily in the third.
4. The aforementioned tabular analysis is tied to the graph by relating the calculation steps (identified by lines a, b, c, etc.) to the step-series graph.
5. In the tabular analysis supporting the step-series graphic presentation, again, meticulous care is taken to identify all steps and figures and to demonstrate what calculations are occurring by means of formulas that utilize the identifications made.

It has been emphasized that to the extent possible, the amount of detail developed for diagrams should be such as to minimize the need to refer to the accompanying text pages. In turn, the extent of this information may depend on the background of the intended reader. Thus there is some room for judgment and divergent opinions. A good rule to follow, however, is that to the extent that space and relative freedom from clutter permit, all detail should appear with the graph or, in this case, with the graph and accompanying table.

As a final point, the table may be viewed as a part of the graph; in the final assembly of the pages of a report or other form of presentation, it may be desirable for the table and graph to appear on opposite pages for ready comparison. The existence of both a graph and a table need not nullify the desirability of explanatory text pages. Complex ideas, concepts, or results often require multifold presentation—graphs, tables, text —for comprehension by all, even though the reader with a fairly good background may be able to skip the text simply refer to the diagrams and accompanying tabular matter.

SPECIAL SCALING

A frequent refinement in graphing is the use of special scaling. As first, many users of graphs may be annoyed by such devices, but their virtue often outweighs initial inconveniences. The most common type of special scaling is "semilog" graphing, with the vertical axis expressed in logarithmic increments. An illustration appears in Figure 4-4, giving regular and semilog scale side by side.

On semilog paper equal percentage increments plot in equal scale steps. For example, a rise in sales volume from $100 to $200 represents a 100 percent increase since ($200-$100)/$100 = 1.0 or 100 percent. Next, from 200 to 300 is only a 50 percent rise. In particular, ($300-$200)/$200 = 0.5 or 50 percent. Therefore, on a log scale, the jump from 200 to 300 is of the same size as the jump from 100 to 150. This can be readily verified from the right-hand plate of the diagram.

Suppose next that we want to project future sales. The left-hand graph is curvilinear, and it is rather difficult to make a further curvilinear extension. Which way should the curve run? On the other hand, the right-hand portion shows an actual linear trend as best fitting the data plotted; it is simple enough to make a further linear extension of this trend. It is, of course, necessary to verify that a logarithmic or, in other words, percentagewise increase in sales volume is actually occurring. But once we have satisfied ourselves of that fact within reasonable limits of confidence, then the log chart helps in extending the projection.

Trends and relationships need not be logarithmic in nature. Logarithms, or the percentages and ratios that they represent, constitute only one type of functional relationship among factors or variables. Usually the analyst working on a report is aware of the type of relationship that underlies his data points, and he will then be able to utilize the appropriate scaled paper. Sometimes, of course, there is doubt about the exact nature of the relationships involved. In such instances, a trial-and-error approach, examining many potential candidates on graph paper, is used until we are satisfied that a "most fitting" scaling has been found.

DUAL SCALING

Special scales permit the use of two separate vertical graduations, one to the left and one to the right of the body of the chart. One type of usage,

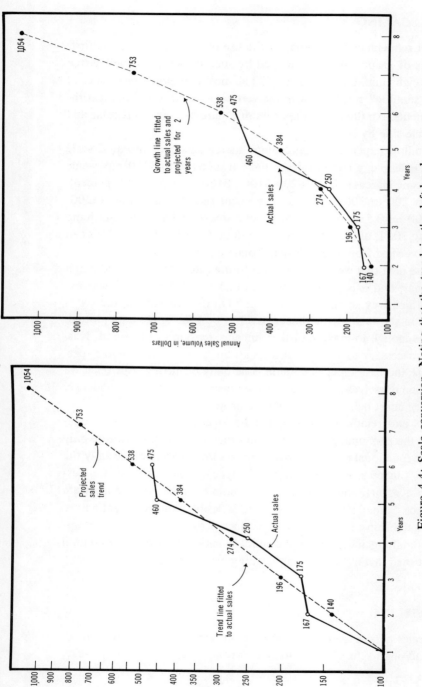

Figure 4-4: Scale conversion. Noting that the trend in the left-hand series represents constant precentage increments, we convert the vertical scale to logarithmic (that is, percentage) increments. A linear trend now results, which usually permits straight-line projections. The prior, non-linear trend is more difficult to handle. Scale conversion, for proper occasions, is thus often quite useful. From N. L. Enrick, *Market and*

in which such a technique is quite popular, is in the graphing of stock market data. Figure 4-5 is a case in point, utilizing semi-logarithmic, or ratio grid, plotting. The left hand accommodates, in logarithmic increments, the earnings and dividend recording of the stock. A different but

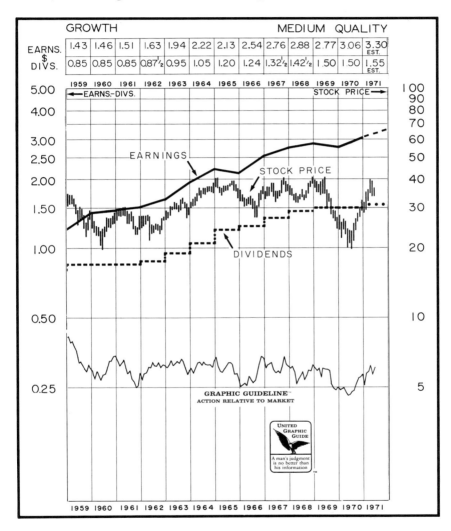

Figure 4-5: Ratio chart with several curves. This example shows prices, earnings, and dividends on a stock (Continental Oil Company), together with an evaluating "guideline" on ratio scalings. From *United Graphic Guide*, vol. 1, p. 50, Boston: United Business Service Company, 1968.

also logarithmic graduation to the right is in terms of the price per share of the stock.

The example, incidentally, represents a relatively compact graphing application. Four curves are plotted, showing past experience and estimated future projection. In addition, a tabulation of detailed numerical values for earnings and dividends per share is given along the top section of the chart. Further, it will be realized that ratio-scaling itself is a means of compressing data in a narrower space than would otherwise be needed.

CORRELATION DIAGRAMS

When relationships between two variables are to be investigated and made explicit, a correlation diagram, as in Figure 4-6, can be most useful. The points are derived from the two data sets for housing starts and plumbing sales of Figure 3-4. But while an examination of the original two time series did indicate some relationship between these two variables, a more specific analysis is possible only through correlation work.

Conversion to correlation is attained by plotting each pair of time series points. For the first year of the two time series, for example, we note that housing starts are at 130, while plumbing sales are at 70. On the correlation diagram, therefore, find 130 on the base scale for "New Housing Starts" and go up to 70 opposite the "Sales Volume" scale, plotting a single point for the prior pair, as shown. Eventually, two points will be drawn in this location, since five years later a similar relation of 130 in housing starts and 70 in plumbing sales occurs.

For the second year of the time series, we have a pair of 115 in housing starts and 60 in sales, again yielding a single point on the correlation diagram, or "scatter plot," as it is sometimes called. Continuing in this manner for all 12 pairs of points, the diagram is filled with the requisite single points.

Next, using mathematical methods, a line of average relationship is computed. The equation of this line is:

$$\text{Sales Volume} = 32 + \tfrac{1}{4} \times \text{Housing Starts}$$

Also known as a "regression line," this generalized indicator of the estimated effect of new housing starts (the independent variable) on sales

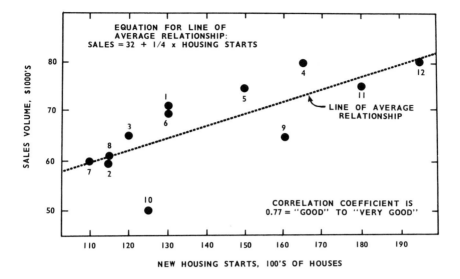

Figure 4-6: Correlation chart. Data from the time series in Figure 3-4 are replotted to bring out more clearly the relation between new housing starts and sales of plumbing. The "line of average relationship" was calculated mathematically to fit all 12 points in the best way. It might have been drawn by free-hand estimate, its quality then depending on the judgment of the draftsman. Figures next to each point show the year number (from year 1 to 12).

volume (the dependent variable) is drawn in on the graph. A plotted point falling above the line indicates sales performance better than the long-term average trend, while conversely, points below the line represent apparently poorer attainments.

MULTIPLE CORRELATION

When more than two variables are involved, we talk of multiple correlation. For example, Figure 4-7 portrays the effect of two input or independent variables in production—particle size and blending percentage —on the dependent or output variable of burst strength of a plastic material. The drawing is in isometric form, with the height of each column representing the strength obtained from various particle sizes and blends tried in actual production.

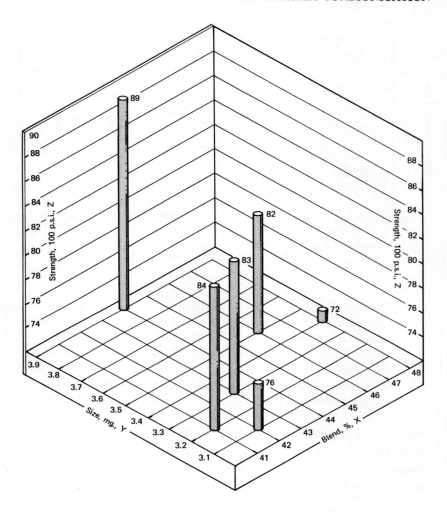

Figure 4-7: Three-dimensional correlation diagram. The effect of blending percentage X and particle size Y jointly on burst strength Z is plotted. A relationship is apparent from this presentation, from high strength at higher size values and lower blending percentages, to low strength at lower sizes and higher blending percentages. From N. L. Enrick and H. E. Mottley, Jr., *Manufacturing Improvement through Experimentation,* Newark, N.J.: General Instrument Corporation, 1968.

SOPHISTICATED CHARTS

For more than three variables, it is usually not feasible to make diagrams; for more than four variables, such a task is a sheer impossibility. Nevertheless, one can mathematically construct a multidimensional model and derive pertinent equations from it. In all instances we can obtain the "correlation coefficient," which is a measure of the degree of relationship demonstrated among the variables. In general, a coefficient of 0.9 shows excellent correlation, while 0.8 is for very good, 0.7 for good, and 0.6 for fair. Values below 0.6 generally represent poor correlation. The coefficient cannot go beyond 1.0 for "excellent." In practice, however, such high value is not expected unless there is some error or other defect in the recording or analysis of the data. For details of calculation of the correlation coefficient, reference should be made to statistical texts.

SYNTHESIS CHARTS

A synthesis chart relates a series of rather general ideas, often in the form of a circular diagram or other symbolic arrangement, so as to construct a new overall design from the component segments. Figure 4-8 is an example. Such a chart is almost never meaningful by itself, but merely accompanies what is usually extensive and explicit text. For our illustration, it is now the author's task to show that he considers the six elements of science—history, content, communications, rationale, analysis, and concepts—as a wheel, specifying the information that requires systemization. The element concept is placed in the center of the wheel, signifying its importance to the development of science. Thus concepts are viewed as the idea area of science, where thoughts are born, tested through analysis, classified according to content, history, and rationale, and communicated. Next, the author shows by means of illustrative examples how all the elements of science interact and how they are synthesized.

There is a similarity between the synthesis chart and the pie chart, but in terms of the circular pattern only, not of the nature and intent of the synthesis chart. If one is to look for similarities, it can be pointed out that the synthesis chart is very much akin to the types of conceptual visualizations and stylized flow charts that will be presented in the next chapter. Indeed, it may often be feasible, and probably desirable, to convert a synthesis chart to a flow chart, since the latter is more versatile and permits more information to be shown.

SCHEDULING CHARTS

For the purpose of short-range planning, various devices known as scheduling charts, or Gantt charts (after their originator), are in use. Figure 4-9 gives an example, showing booked manufacturing orders, actual progress of production, and the "today" line. Thus, how far we are booked for production, whether or not actual manufacturing is ahead or behind of schedule, and related information of interest to management is readily visualized.

When more complex and long-range projects are involved, the Gantt chart is amplified to a PERT network, as shown in Figure 4-10. Here we

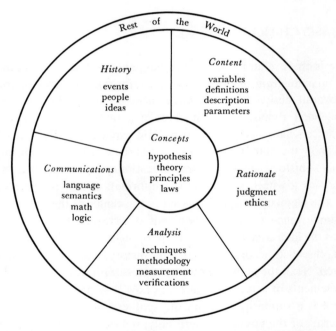

Figure 4-8: Synthesis chart, which results from observing a phenomenon (such as the development of science) in terms of its elements (history, content, communications, rationale, analysis, and concepts). Highly abstract in nature, a synthesis chart usually requires considerable text-ual support, justifying the manner in which individual elements have been synthesized into the new construct. From C. Glenn Walters, "The Order in Marketing Science," in *Business Perspectives,* vol. 4, no. 2 (Winter 1968), p. 33.

add information revealing the interdependencies among events. For example, the graph shows that operation no. 7 cannot be initiated until operations 2 and 3 have been completed. Scheduling charts are a type of flow chart, which will be taken up in the next chapter.

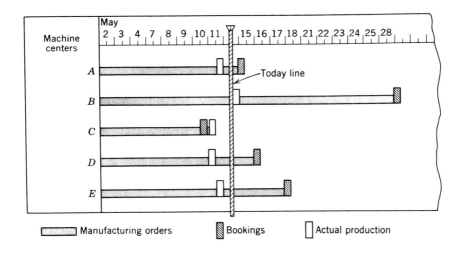

Figure 4-9: Gantt chart. This chart is used for scheduling. Production is assigned to various machine centers and the status of actual production compared with bookings and the "today" line. Note, for example, that output on machine center E is a half-day behind schedule.

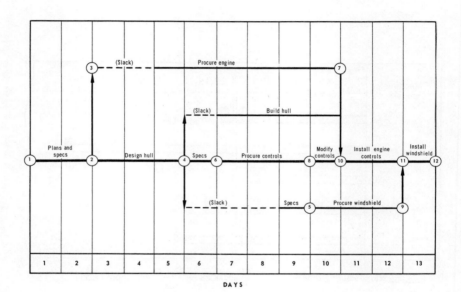

DAYS

Figure 4-10: PERT network. An expanded Gantt chart, this diagram connects planned activities. In this example, it is for the building of a boat. "PERT" stands for "program evaluation and review technique," a system in which this type of scheduling diagram is utilized. From A. Kusner, P. A. Doherty, and J. G. F. Wollaston, "Basic Concepts and Techniques of Pert," New York: Cresap, McCormick and Paget, April 1964, 30 pp.

5. Flow Charts

The discussion thus far has centered on graphs that effectively present numerical data and related items of information. Another highly useful application of visual aids is the representation of various types of flow in chart form. Material that may seem abstract in textual form can often be given clarity and life when presented visually. In turn, a writer, in attempting to convert his thoughts into flow diagrams, may discover and close gaps or other defects in his original set of concepts. A flow chart is thus not only an aid to the reader, but also a medium whereby the writer can clarify, check, and develop his own ideas or tighten the logic of his argument.

PROCESS FLOW CHARTS

The typical chart in Figure 5-1 is a schematic of the physical transfer of materials from one production stage to the next. This particular illustration shows how a bale of man-made staple fibers is opened, blended, and then put through successive operations until the final spinning and spooling has occurred. We are concerned here with the physical flow. The drawings depicting each machine add interest to the chart, but are not necessary; boxes indicating each operation would have sufficed.

SYSTEMS OPERATION CHARTS

An extension of the methodology of process flow charting leads to systems operation charts such as is depicted in Figure 5-2. The "process" has become the operation of a system. The particular system represents a pro-

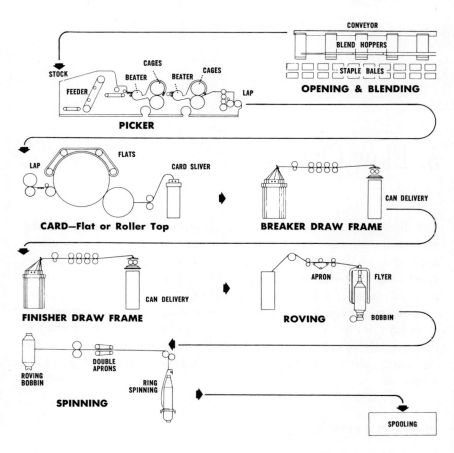

Figure 5-1: Process flow chart. Physical progress of materials is illustrated in this opening-through-spinning application. Instead of physical processes, flow charts may also depict flow of information, procedures, decisions, or other more abstract conceptual processes. From Chemstrand Division, "Blanket Manufacture: Technical Information on Acrilan A-15," July 1966, Monsanto, Decatur, Ala.

gram for integrated quality control, from raw materials through production and distribution. Feedback is provided, whereby consumer reaction is gauged through consumer and market research, leading to product design and redesign, supplemented by process engineering. Continual improvements in satisfying consumer requirements are thus sought. The diagram, in a nutshell, shows the essential operational aspects of the system.

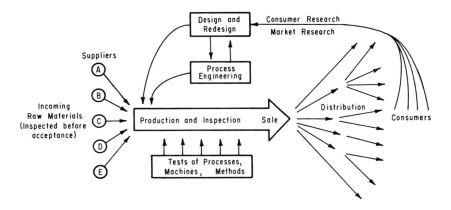

Figure 5-2: Systems operation chart. The illustration covers an integrated quality control system, from incoming materials through distribution, market research, product design and redesign in the light of this research, and concomitant process engineering work. From W. Edwards Deming, "Management's Uses of Statistical Methods," in N. L. Enrick, *Cases in Management Statistics,* New York: Holt, Rinehart & Winston, 1962.

The example is of a relatively simple chart. If examined in detail, systems are usually more complex. Often, of course, it is not possible to represent an entire system on one graph. In such instances, an overall view can be given, with the provision that various segments of the chart are shown in separate graphs. For our particular example, the interaction between product design, redesign, and process engineering may be a very complex one, involving many rechecks of consumer and market research data, availability of raw materials supplies, and production feasibilities. A separate chart can be utilized to show these functions in detail. Other aspects, such as the relation of processing, machine and methods analyses, and tests of production and inspection, can be treated similarly by separate charting. The systems operation chart gives a quick overview of the total system, supplemented where necessary by individual subsystem operation charts.

As a further illustration, Figure 5-3 shows a flow chart concerned with an idealized problem-solving sequence. Again, there is a feedback mechanism. In fact, superficially the diagram looks similar to the preceding one, although in reality they are quite different. The problem-solving sequence portrays a far more generalized system than that of the prior diagram.

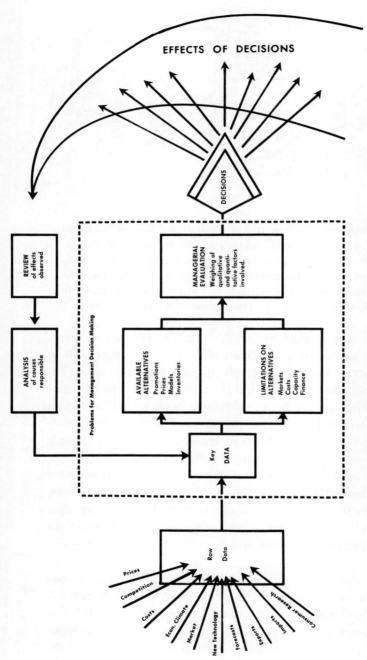

Figure 5-3: A further illustration of the system operation chart. The ideas for patterning the flows obviously came from the similar design of the diagram in Figure 5-2. The subject matter of the two charts is nevertheless quite different. Entirely unrelated material, topic-wise, can thus be the source idea for the development of new structures, concepts, or other creative combinations. From N. L. Enrick, *Inventory Management*, Scranton, Pa.: © Chandler Publishing Co., a division of Intext, 1968.

PROCEDURAL FLOW CHARTS

A chart closely related to the type just shown is the procedural flow chart, such as that given in Figure 5-4, illustrating the basis of gear-change orders issued in a plant. Aside from clarifying the requirements involved in the changing of gears, thereby controlling their issuance and usage on manufacturing equipment, the chart also aids the designers of the procedures. In this particular instance, the initial procedural flow chart revealed gaps and loopholes that were corrected in the final version.

Procedural flow charts emphasize an interesting factor in the problems of management. The activities of several people must be meshed well, leaving no serious gaps or potentials for error, so that the total system will work as intended.

Procedural flow charts may also become quite generalized, as is illustrated in Figure 5-5, portraying the steps pursued by a management operations research team in tackling managerial problems with the aid of computer methodology. Although the real-life situation undoubtedly involves some feedback mechanisms, the chart fails to portray it. The

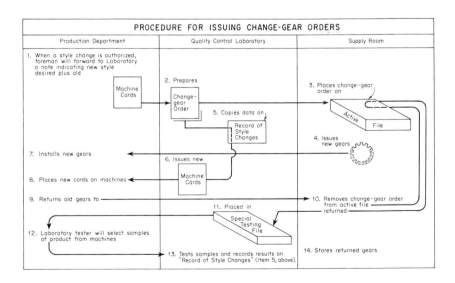

Figure 5-4: Procedural flow chart, an example of methods to be used by production, quality control, and supply room in the initiation, issuance, and installation of gear changes in a manufacturing plant.

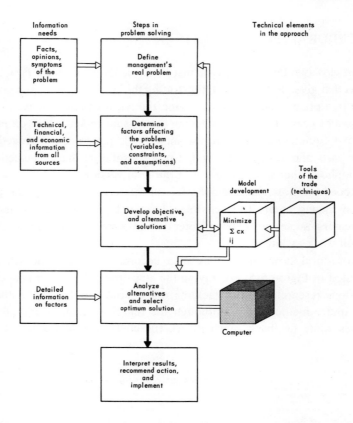

Figure 5-5: Procedural flow chart, depicting the steps followed by an operations research team in solving managerial problems with the aid of a computer. From Robert A. Hammond, "Making Operations Research Effective for Management," in N. L. Enrick, *Management Operations Research,* New York: Holt, Rinehart & Winston, 1964.

author may well have felt that this aspect was of relatively minor significance in the context of the principal steps to be portrayed and that its inclusion would have caused needless clutter.

Some readers may rightly feel that the differences between a procedural flow chart and a systems operation chart are relatively difficult to establish, since procedures are generally conceived and designed with a system in view. Admittedly, the distinction may be a fine one, and individuals' views may overlap. Uses made in this book are matters of convenience, without imputing any general standards of definitions.

SEGMENTATION OF FLOW CHARTS

The two procedural flow charts just discussed have an interesting feature: there is a columnar segmentation. For example, our last chart, showing the steps followed by an operations research team in solving managerial problems, is segmented under three headings:

1. Information needs
2. Steps in problem solving
3. Technical elements involved in the approach

As the procedural steps flow through their various stages, they are characterized by the type of information, method, or operation involved.

The chart showing the method to be followed in initiating and executing change-gear orders on machines also has a segmentation. This time, the sectioning is in terms of the operating departments involved, consisting of production, quality control, and supply room.

Whenever it is possible, one should segment procedural flow charts in terms of the major categories of operations involved. It will be noted that this prescription was not followed for the systems operation charts— for a good reason. When many departments and a multitude of types of operation are involved, it is no longer feasible within the space of a simple chart to compartmentalize along generalized lines.

CONCEPTUAL VISUALIZATION

A flow chart may become quite abstract, representing merely the relationships among principal concepts, factors, and ideas that must be considered in the context of a problem. Figure 5-6 gives such a case. We are concerned here with a diagram that serves to illustrate the relation of various concepts that, together, form a model of a complex system development process. Arrows show the flow of data and analysis procedures, together with provision for feedback via the "system model" loop.

The flow shown is of a very general nature. One may view the boxes for "Measure of effectiveness," "Systems goals and requirements," and so on as the major factors to be considered in evaluating the effectiveness of a system in performing a certain critical activity. Such items as "Need

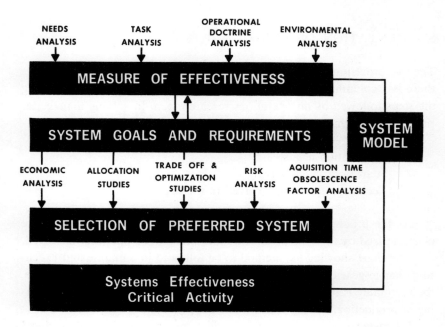

Figure 5-6: Conceptual visualization, illustrating the relation of various concepts that together form a model of a complex system development process. Arrows show the flow of data and analysis procedures. The "system model" is used for testing and feedback of information. From P. J. Giordano, "Predicting Systems Effectiveness," in "Proceedings of the Second Naval Material Support Establishment System Performance Effectiveness Conference," April 1966, U.S. Naval Applied Science Laboratory, Systems Effectiveness Branch, Brooklyn, N.Y.

analysis," "Trade off and optimization studies," and so forth represent the analysis tools used in evaluating the major "boxed" items. Certain interchanges such as the relation between measures of effectiveness and the goals of the system are indicated.

A general word of caution applies to all conceptual visualizations. While it is true that this is often the only practical way to present a general view of a relatively intricate configuration among concepts, ideas, analyses, and considerations, one should make certain that greater precision is not attainable. Certainly, a systems operation chart or a procedural flow chart represents successively higher degrees of precision. It thereby indicates just what is happening either in the "real" or "conceptual" world that is to be reflected by the chart. Therefore, when using a conceptual visualization, we must ask ourselves:

1. Could this be converted to a systems operation chart, and if not, then,
2. Could it be revised to reveal more of the actual flow of related factors under consideration?

Sometimes a change from a conceptual visualization model may proceed first to the systems operation chart and from there to the still more precise form of a procedural flow chart. In any case, one will usually need to draw and redraw a chart several times until a truly useful end result has been accomplished. The process is intellectually quite interest-

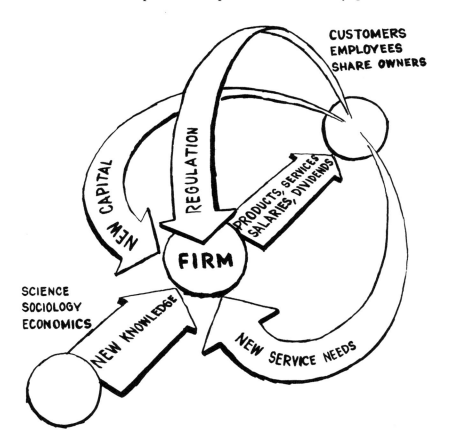

Figure 5-7: Stylized flow chart, showing the functions of a corporation and its relations to the business environment, in stylized outline. From Jack A. Morton, "The Manager's Changing Role in Technological Innovation," *Bell Telephone Magazine*, vol. 47, no. 1 (Jan.-Feb. 1968), pp. 8-15.

ing. The successive designing and redesigning of the visual representation of an author's ideas results in his own rethinking, reanalyzing, and re-visualizing of the original concepts. With this enhanced understanding comes increased clarity and ultimately a better chart.

STYLIZED FORMS

A stylized diagram is even more generalized than a conceptual visualiza-tion and is thus similar to a work of art. Figure 5-7 is an example which explores in broad general outline the relation of a corporation to its en-vironment. The impact of such a diagram on reader consciousness may be considerable, however. A telling point, thus underscored visually, is likely to sink in and be remembered longer.

Stylized forms are often used to emphasize the principal message of an article, report, or book. An example of such a chart, imbedded in the jacket of a book, appears in Figure 5-8. When used in this manner, no chart title or specific explantory text need be given, since the message or idea of the diagram should be self-evident or else easily gleaned from material surrounding it.

CONVERGENCE CHARTS

A series of separate flows may converge, such as when a large set of detailed data lead to an overall result. For example, Figure 5-9 shows how business transactions data leads to a few financial figures which ulti-mately yield "return on investment" as the end result of the sequence.

We may say that the individual data converge toward a single end result, hence the name "convergence chart." Another view holds that one item—"return on investment," in this example—is built on a stream of prior figures. The diagram, under that position, would be characterized as a "build-up chart."

PROPORTIONED FLOW CHARTS

When the relative magnitude of individual flows is shown by the size of the stream, we have a proportioned flow chart, such as is depicted in Figure 5-10. Arrows within the unshaded bands indicate main flows and

Figure 5-8: Stylized flow chart appearing on a book jacket. In very general terms, the chart is to convey the idea of systematic planning involving men, machinery, plant, and distribution. From N. L. Enrick, *Management Planning,* New York: McGraw-Hill Book Co., 1967.

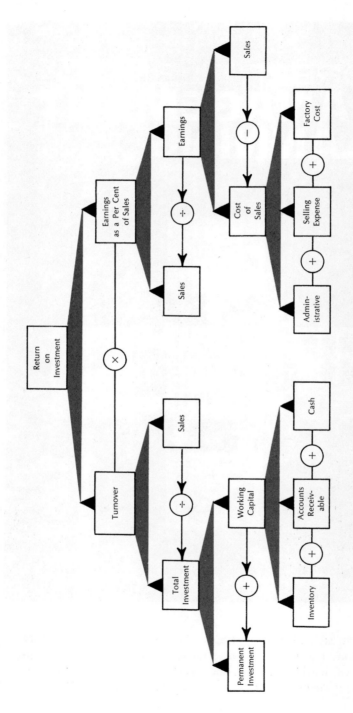

Figure 5-9: Convergence chart. The convergent flow of financial data, leading to return on investment, is portrayed in this application. Flow of entries and types of calculation (addition, subtraction, multiplication, and division) are indicated.

cross-flows. The width of each band is in proportion to the relative magnitude of each main flow and cross-flow of the monetary funds involved.

DEPENDENCY CHAINS

A dependency chain such as that depicted in Figure 5-11 is similar to a flow chart. It portrays a sequence of events. But instead of emphasizing the flow, it concentrates on the dependence of various events on each other.

1972 Budget – Relation of Budget Authority to Outlays
Figures in brackets represent Federal funds only

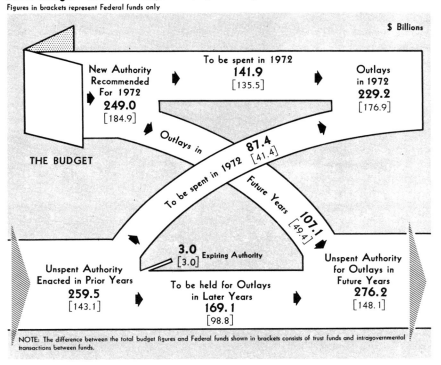

Figure 5-10: Proportioned flow chart. The illustration comes from the federal budget. Expenditures authorized in one year are not all spent in the same year, thus giving rise to carryover of outlays to future years and a backlog of unspent authorizations from prior years. The resultant flows are depicted. From "The Budget in Brief," Executive Office of the President, Bureau of the Budget, U.S. Government Printing Office, Washington, D.C., 1971.

In actual applications there are many examples of the use of dependency chains. For example, in scheduling production over a series of processing stages, the operations at one stage depend on completion of other operations at prior stages. In research and development work, similar dependencies in the invention of various devices or systems, forming subphases of the final project, have intricate configurations.

An example of a time-dependent chain is given in Figure 5-12, which traces the travel of orders and merchandise between retailer, wholesaler, and factory. The same information in more detailed form appears in Figure 5-13. Finally, when the results of these last two graphs are plotted in time series sequence, as in Figure 5-14, the relationship of this type of presentation to dependency chains becomes clear. The last three diagrams also serve to reiterate a point previously made: sometimes it is desirable to build up an idea, concept, or result in step-wise fashion. For this purpose, several charts may be needed.

Figure 5-11: The box represents an event or action that takes place. The heavy circle represents a component part; the triangle shows an event that must operate, so that the item to the right of the triangle can function. For example, unless the weight in *K* falls, the lever at *O* will not trip the current.

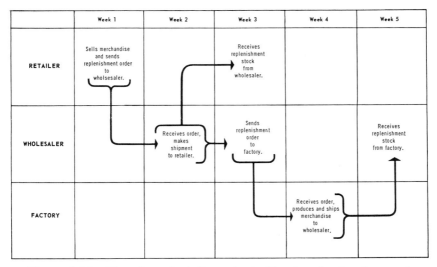

Figure 5-12: Time-dependent flow chart. The travel of orders and merchandise among retailer, wholesaler, and factory is depicted graphically. The transactions from one week to the next make this a time-dependent chart. From Enrick, *Market and Sales Forecasting,* Scranton, Pa.: © Chandler Publishing Co., a division of Intext, 1969.

As a final illustration of a dependency chain, Figure 5-15 shows the interdependency of various courses and the time at which they should be taken, in relation to a college curriculum leading to the Master of Business Administration degree. The author utilizes the diagram to accomplish several objectives:

1. Show the time (in terms of year and semester) at which each course should be taken.
2. Describe the general nature of each group of courses, such as material concerned with functions, tools, environments of business, and personal development.
3. Develop the additional concepts built from these foundations, in terms of organization, operation, and control of a business, and show how they serve in completing a "business enterprise" workshop.

The graph emphasizes time-dependence. The functional interdependence of the courses in relation to the concepts shown in the arrow section of the graph is not developed specifically. A supplemental graph, showing

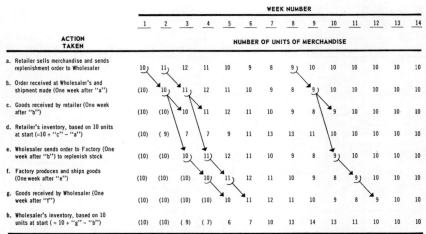

	1	2	3	4	5	6	7	8	9	10	11	12	13	14
ACTION TAKEN	WEEK NUMBER — NUMBER OF UNITS OF MERCHANDISE													
a. Retailer sells merchandise and sends replenishment order to Wholesaler	10)	11)	12	11	10	9	8	9)	10	10	10	10	10	10
b. Order received at Wholesaler's and shipment made (One week after "a")	(10)	10)	11)	12	11	10	9	8	9)	10	10	10	10	10
c. Goods received by retailer (One week after "b")	(10)	(10)	10	11	12	11	10	9	8	9	10	10	10	10
d. Retailer's inventory, based on 10 units at start (=10 + "c" – "a")	(10)	(9)	7	7	9	11	13	13	11	10	10	10	10	10
e. Wholesaler sends order to Factory (One week after "b") to replenish stock	(10)	(10)	10)	11)	12	11	10	9	8	9)	10	10	10	10
f. Factory produces and ships goods (One week after "e")	(10)	(10)	(10)	10)	11)	12	11	10	9	8	9)	10	10	10
g. Goods received by Wholesaler (One week after "f")	(10)	(10)	(10)	(10)	10	11	12	11	10	9	8	9	10	10
h. Wholesaler's inventory, based on 10 units at start (= 10 + "g" – "b")	(10)	(10)	(9)	(7)	6	7	10	13	14	13	11	10	10	10

Arrows indicate direction of time flow. Parentheses denote entries from Sales in prior weeks (before Week No. 1) with stable demand at 10 units per week

Figure 5-13: A further application of the time-dependent flow chart. The flow of orders and merchandise between retailer, wholesaler, and factory, previously shown in entirely schematic form, is now supplemented in terms of a more complete format. A larger number of weeks and a detailed explanation of quantity flows are provided. From Enrick, *Market and Sales Forecasting,* Scranton, Pa.: © Chandler Publishing Co., a division of Intext, 1969.

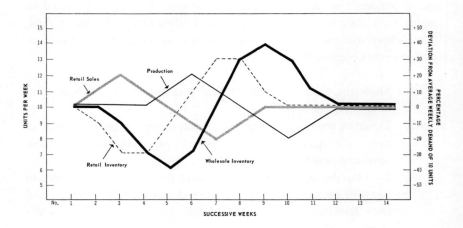

Figure 5-14: Time series presentation of time-dependent flow chart data. The plots are derived from the flows shown in Figure 5-13. The graph here gives further meaning to the results of the analysis. From Enrick, *Market and Sales Forecasting,* Scranton, Pa.: © Chandler Publishing Co., a division of Intext, 1969.

82

these additional aspects of interrelationships, could of course be prepared. As a further idea, one might try to develop a graph that simultaneously traces the links of the dependency chain as regards both time and course-content factors.

ORGANIZATIONAL PATTERNS

Generally the relationships and flows of authority and responsibility in an organization are depicted by means of schematics such as the three diagrams in Figure 5-16 for various types of industries. A more advanced application involves the development of organizational patterns that (1) separate "authority and responsibility" lines from "confer and advise" connectives by means of solid and dotted lines, and (2) indicate the type of responsibility, such as that denoted by the segments designated A, B, and C in Figure 5-17.

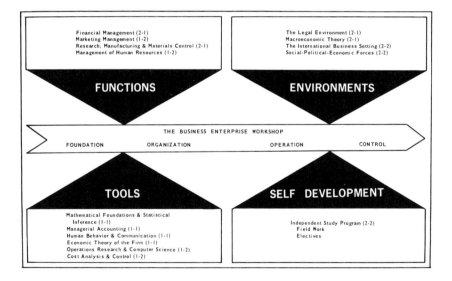

Figure 5-15: A dependency chain in schematic form. This example represents the program and underlying philosophical framework of a college curriculum for the Master of Business Administration program (at the University of Notre Dame). Reproduced with permission from *Collegiate News and Views,* December 1967, published by Southwestern Publishing Co.

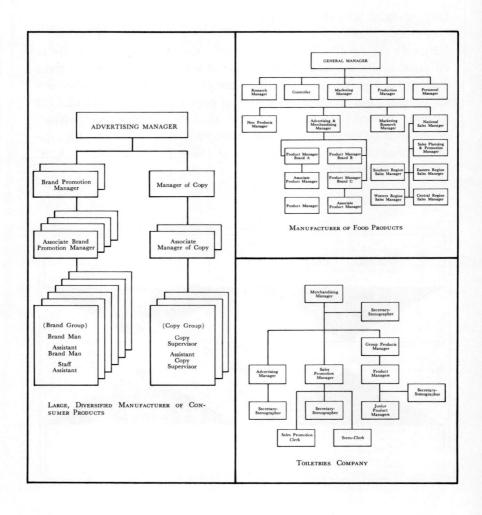

Figure 5-16: Organization charts showing flow of authority and re-sponsibility. From Robert M. Fulmer, "Does the Product Manager Manage?", in *Business Perspectives,* vol. 4, no. 2 (Winter 1968), p. 12.

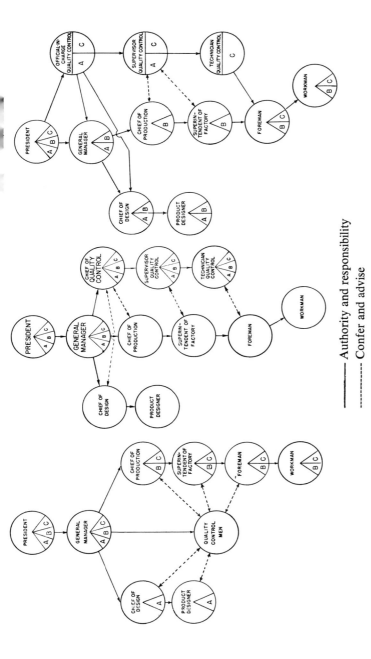

Figure 5-17: Organizational patterns. Three possible ways in which quality control may be structured within an organization are shown in this example: (a) on an advisory basis, (b) in a departmentalized manner, and (c) as a function of top management. Letters A, B, and C indicate various types of responsibility. A is for quality standards, B for product quality, and C for determining and reporting quality of product. From N. L. Enrick, *Quality Control and Reliability*, 5th ed., New York: Industrial Press, 1966.

85

6. Decision Charts

A widely recognized axiom of management states that "well informed is well decided." Graphic materials that present worthwhile information lucidly are thus useful aids to the person or group who must make decisions. The term "decision charts," however, is reserved more specifically for specialized types of visual materials that address themselves directly to decision functions.

BINARY DECISION FLOWS

When electronic computers were introduced, they functioned on a binary flows system. This means that a tube, transistor, switch, circuit, or other unit could be in only one of two stages, such as "on" or "off." This binary principle can support a highly advanced and intricate automated data processing system. There was a concomitant need to show the data and information flows, as programmed on a computer, in visual form. The type of chart developed for this purpose again relied on a binary flows scheme, an example of which will be given later in this chapter, after the principles of the technique have been explored more fully.

We have noted that when a unit, system, or other item can be in either one of two possible conditions, a binary process is operative. It should be added that the condition need not be "on" or "off," as for a switch. Other binaries are "yes" or "no," "applicable" or "inapplicable," and so on, as long as either of the two states is exclusive, in the sense of not admitting a third condition, state, action, or operation.

BINARY DECISION CHARTS

Probably the best way to study a binary decision chart is to look at an example, such as the "How to Get Up in the Morning" diagram in Figure 6-1. In this humorous illustration, observe the following:

- Events representing actions are shown in rectangular boxes.
- Points at which a question is posed are diamond-shaped.
- Each question is stated so as to call for one of two answers, "yes" or "no."
- Depending on the answer given, one of two decisions is recommended or prescribed.
- Feedback is possible and is generally denoted by arrows that move counterclockwise. One such feedback situation, between "groan" and "shake wife," is indicated on the diagram.

Points could be made regarding the amount of detail shown and some argument could be advanced for developing a larger, more explicit chart for most any decision-action sequence. But we must always keep in mind that too much detail in the way of minutiae or other particularization will detract from the information content as a whole. Oversimplification must also be guarded against. It would thus be highly inappropriate to make a chart that merely says: "1. Get up, 2. Get ready, 3. Go." Ultimately, it is the purpose and intent of the decision chart that determines the appropriate degree of particularization called for.

We shall present several practical examples of decision charts, but first another basic point needs clarification.

MULTIPLE-CHOICE DECISIONS

Most decisions in management, science, technology, and life in general involve more than just a binary choice. Computer designers and decision chartists are well aware of this and have developed a system of successive binary build-ups so that multiple-choice questions can be handled on the basis of the simple on-off binary unit. Confirmation of this extensibility is given in Figure 6-2, which enlarges on a segment of Figure 6-1. Three kinds of decisions are now possible: (1) kiss and embrace wife, (2) kiss her briefly, and (3) forget to kiss her. These actions, in turn, depend on

HOW TO GET UP IN THE MORNING

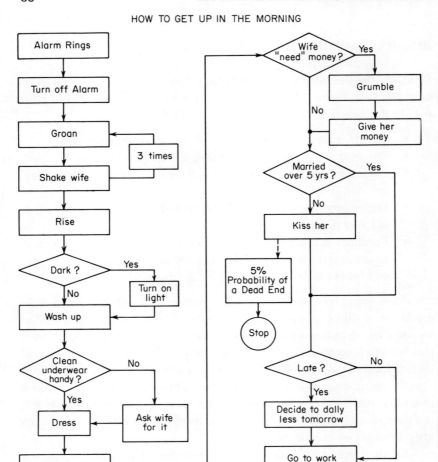

Figure 6-1: Binary decision flow chart. Items in diamonds contain questions with "yes" or "no" answers and thus call for one of two decisions.

the threefold question of whether the couple is married (1) less than a year, (2) between one and five years, or (3) more than five years.

What device has been used to answer a threefold decision problem by means of twofold binaries? An examination of the example just given yields the answer. By building a system of two successive binary question diamonds, we have come up with a dual binary pattern that provides decisions to a threefold question. Extensions of fourfold, fivefold, and multifold systems are accomplished simply by enlarging the pattern of successive binary sequences.

Finally, it will often be found that designers do not bother to show a dual or otherwise multiple build-up in the detail we have just used. To handle the problem in our example, they would simply use a diamond with the question, "Married how long?" Next, they would use the three unused corners of the diamond to say "less than 1," "1 to 5," and "over 5" on three lines that lead to the corresponding decisions. A dual binary system is thus implied though not actually articulated.

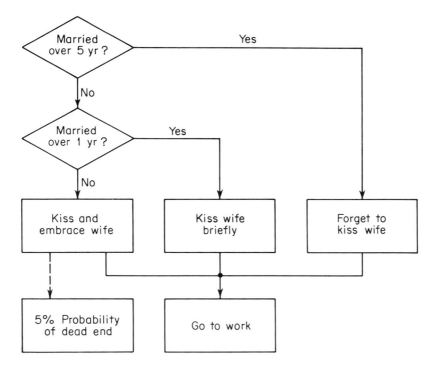

Figure 6-2: Binary decision flow chart for multiple-choice questions. This illustration shows how a sequence of two binary questions permits a decision to three questions: (1) "married over 5 years?", (2) "married less than 5 years but over 12 months?", and (3) "married less than 12 months?". The decision made (or outcome) depends on which question applies at a given point. By extending the sequence of binary questions, we can pose as many questions as are needed. (Note that the illustration here is an extension of Figure 6-1.)

MANAGERIAL APPLICATIONS

It is time to look at practical uses of binary decision flow charts. The diagram in Figure 6-3 comes from a marketing experiment application. A new product has been conceived by the firm's product development department and a series of steps called "marketing experimentation" are

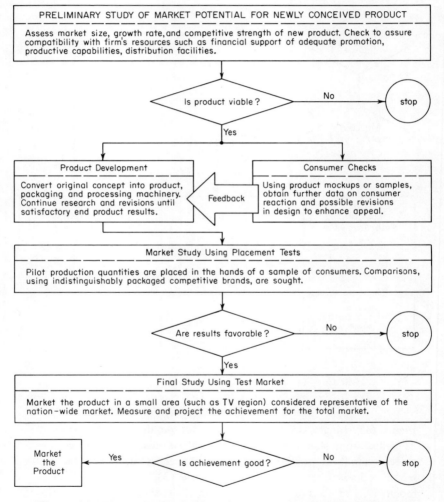

Figure 6-3: Binary decision flow chart. The illustration shows the "new product development cycle" from conception to consumer. From Enrick, *Market and Sales Forecasting,* Scranton, Pa.: © Chandler Publishing Co., a division of Intext, 1969.

now required. Only if the results of this work are satisfactory can one justify the tremendous costs and often not inconsiderable risks involved in the full-scale promotion, marketing, and distribution of the product. The series of analysis steps involved should be self-evident from a perusal of the diagram. Note the consistent use of rectangles, diamonds, and circles for actions, questions, and "stop," respectively.

A further illustration is given in Figure 6-4, relating to a managerial problem-solving sequence, with special emphasis on two alternative goals of "satisficing" versus "optimizing." While the questions at each stage remain the same for either of the two goals, the resultant action types vary. The example shows how a dichotomy—two distinct goals in this instance—can be carried along in a decision flow chart.

The illustration points up an interesting multiple use of feedback. Although not stated specifically, when feedback from one point leads back to three different types of action, it is implied that the closest action box will be tried first; in case this should not yield the desired or required results, succesive boxes then become applicable.

MATHEMATICAL APPLICATIONS

The versatility and wide applicability of binary flows charts is further demonstrated by the illustration in Figure 6-5, which is a mathematics-oriented usage, giving the steps involved in the evaluation and solution of a system of linear equations. While the procedures described are somewhat technical, the wording used and the language and symbols of the chart are adequate in relation to the intended audience. Even without being a mathematician, one will note that the decisions to be made at various stages of the analysis depend on comparisons of results obtained at successive points. An intensive feedback operation is shown. In fact, the problem solution is portrayed near the center of the procedures flows, as the ultimate outcome of two sets of feedback cycles.

COMPUTER PROGRAMMING APPLICATIONS

When computerized systems are developed, flow binaries are needed not only for program design, but also to communicate to all concerned the manner in which the system is to operate. Figure 6-6 shows a relatively

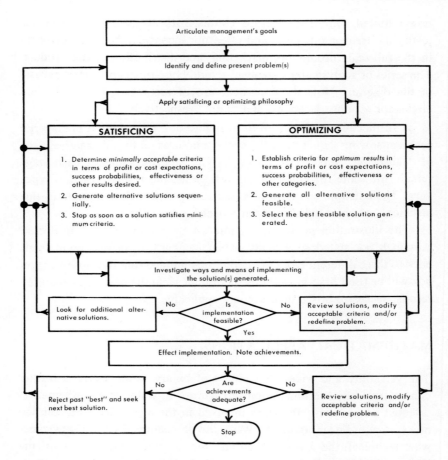

Figure 6-4: Managerial problem-solving sequence. After management's goals have been articulated and present problems identified or defined, the process of seeking a solution depends on whether a satisficing or an optimizing philosophy is pursued. Feedback—involving review, redefinition, and adaptive changes—occurs when (1) alternatives are found unfeasible during implementation, (2) results attained fail to live up to anticipations.

simple and confined application, designed merely to demonstrate the general nature of such charts. Unfortunately, most of the diagrams depicting computer flows must contain so much material within a limited space that a telegraphic style of writing, often with extreme abbreviating, becomes necessary. As a result, even a programmer may be unable to fully interpret someone else's charts—unless there is verbal supplementation or a good deal of explanatory material in separate pages of text.

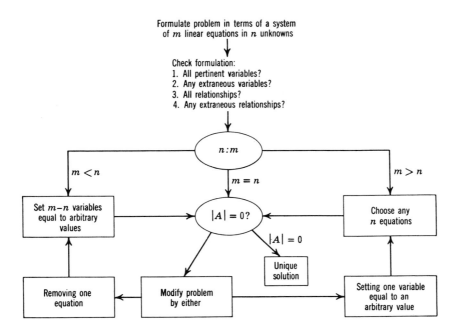

Figure 6-5: Binary chart for a mathematical program. Steps in solving a system of linear equations are developed. The first oval with *"n : m"* says "compare *n* and *m* and determine which is larger. The second oval is a question, but the author prefers not to use a diamond. Note further that the first oval is really an implied question on whose outcome one of two actions applies. The "A" in the center stands for "determinant" of the system. From Daniel Teichroew, *Introduction to Management Science: Deterministic Models,* New York: John Wiley & Sons, 1964.

TREE DIAGRAM

A configuration such as that in Figure 6-7 is commonly referred to as a tree diagram or decision tree because of the dendrite flow or branching out of the guide lines in the chart. The example comes from a project decision problem. It involves these aspects:

Phase 1: We (that is, management of "our" firm) must decide on whether or not to pursue a new project, such as doing research and development work on a new type of material, product, process, or machinery. The chances of success are estimated at 80 percent, or 0.8 in decimal form. Correspondingly, there is an 0.2 chance of failure—that is, we

will do the research and development work, but the ultimate result is that all efforts do not yield a really market-
able material, product, or other desired item.

Phase 2. Regardless of whether or not our firm succeeds in the research and development, a competitor who is working on
his own project is estimated to have an 0.7 chance of success and 0.3 chance of failure. The possible outcomes are
now four: (1) we succeed, he succeeds; (2) we succeed, he
fails; (3) we fail, he succeeds, and (4) we fail, he fails.

Phase 3: Working along to this stage, we note that there are various
chances for a good, medium, or poor market for the ultimate product. Twelve possible combinations must now be
considered.

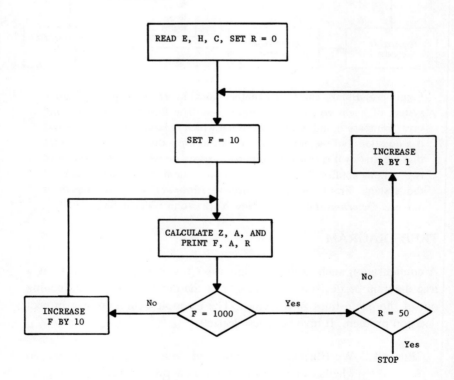

Figure 6-6: Binary computer program. In this example is calculated the
steady state alternating current for a frequency range F of 10 to 1000
cycles per second and resistance R from 0 to 50 ohms. Symbols, A, C,
E, and Z indicate current, capacitance, volt, and impedance, respectively.

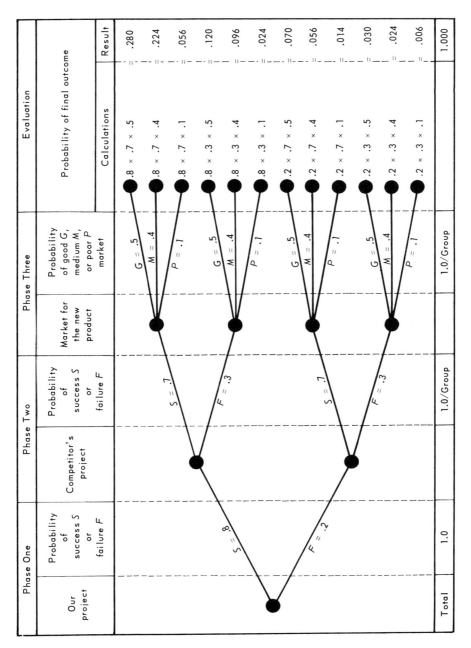

Figure 6-7: Three phases of a program, together with their possible outcomes and probabilities of occurrence, are shown. The final probabilities for each of the 12 possible combinations of paths along the tree may be combined with cost and profitability values to anticipate expected ultimate values. Management's decision, to undertake or drop the project under consideration, is then aided by the expected values calculated. From N. L. Enrick, *Management Planning,* New York: McGraw-Hill Book Company, 1967.

Multiplying the various probabilities together, as shown in the evaluation column, yields combined probabilities for each of the 12 possible final outcomes. The tree is complete. The probabilities may now be used in conjunction with estimated cost and profitability data to obtain the estimated value of the project, considering all of its three anticipated phases. From these evaluations, data result from which management of "our" firm is now enabled to decide on the wisdom of either starting or else dropping the project.

In the example, phases 2 and 3 represent relatively passive states. Outcomes depend on the likelihood of certain competitive actions and probabilities of various types of markets. Another form of tree diagram, representing essentially parallel configurations, involves a large number of phases, some passive and others permitting intermediate management decisions. Decision trees have been gaining increasing adoption as aids in executive actions. Unfortunately, for a great many problems, the choices of action and the possible outcomes are so numerous that bulky and complex trees result. It will be realized that in situations of this nature an actual diagram may not be used. Instead, the phases, pathways, and probabilities are programmed into a computer, which refers to additional cost and profitability data in arriving at ultimate comparative value expectations and estimates, which are then utilized by the decision-makers concerned.

By addressing itself directly to the problem of estimating expected value, so as to permit choice of an optimal course of action, the tree diagram is an excellent decision-furthering tool. It is increasing in popularity particularly wherever large risks and considerable degrees of uncertainty are involved.

7. Charting Grids

In this chapter the types of grids most frequently needed for charting are discussed.

GENERAL PURPOSE DESIGN GRIDS

Shown in Figure 7-1, the general purpose design grid is useful for planning almost any kind of graph. There are five lines to the inch, both ways. Every tenth line is heavy for easier reference. Some space is left in the margin to allow for scaling and identifying text.

TYPEWRITER GRIDS

The typewriter grid in Figure 7-2 allows the planning of graphs and tables that conform to typewriter spacing. The reader can design a chart and leave appropriate space that will permit a typist to insert such text as accompanies the graph. There are 12 vertical and 6 horizontal lines to the inch. Again, every tenth line is heavy, thereby facilitating the use of the grid in designing the charts. For horizontal pages of material, Figure 7-3, with identical grid type, is useful.

A supplementary, useful application of this grid is in designing information and questionnaire forms that conform to standard typing space requirements.

No work is dcne on the grid itself. The grid is used as a basic framework, with a thin bond paper overlaid to permit the grid to show through. One sheet of grid paper thus serves for as many charts as may be desired.

ISOMETRIC GRIDS

The term "isometric grid" refers to a paper such as that shown in Figure
7-4. The cross-hatching effect is useful in providing guidelines for three-
dimensional illustrations.

In order to use this type of grid, the reader may first seek to draw
various size cubes with the aid of this paper. The next step is to duplicate
the three-dimensional figures given in Chapter 4. With this experience,
the average person will be prepared to plot his own three-dimensional
diagrams.

OTHER GRIDS

A multitude of special grids, such as various types of logarithmic paper
(see Figures 7-5 through 7-11), are occasionally required. Other specialized scaling, which is available commercially, may at times be needed.
There is, for example, Weibull paper and Normal Probability paper, permitting plots in terms ot Weibull and normal-curve scalings. It is beyond
the scope of this book to provide these special-purpose grids. When the
need for them arises, suitable paper printed by a variety of supply houses
can be obtained from larger stationery stores. For most purposes, the grids
provided in this chapter are the ones most often used.

OUTLINE MAPS

Outline maps, such as the one for the United States shown in Figure
7-12, represent still another type of charting aid. Among its many uses
are the following:

- Showing population densities by various shadings.
- Indicating size and location of warehouses.
- Tracing routes, such as truck routes or salemen's tours.
- Entering ratios, such as the ratio of sales to population density.

Combination uses are possible. For example, the types of columns
previously presented in connection with bar charts can also be used on

outline maps, such as to show the relative magnitude of ratios and other values applying to each state or to a region.

DRAFTING TEMPLATES

Drafting templates, such as those pictured in Figures 7-13 and 7-14, are of considerable practical usefulness. Cut-outs permit consistent symbols and spacings to be drawn on charts quickly.

PROGRAM AND SYSTEMS SYMBOLS

When routine applications of flow charting of computer programs and operating systems is common, it is desirable to have an agreed upon standard to identify various repetitive types of procedures. International Business Machines Corporation has developed a standardized set of program flowchart symbols and system flowchart symbols (Figure 7-15). In general, detailed information is to be written within the confines of the boxes, diamonds, circles, triangles, and other shapes provided by the symbols.

The system recommended and used by IBM has found wide (but not universal) application. There are variations. Also, while IBM provides a template containing the various symbols, in practice one may require larger borders than those permitted by the template. Also, unless the audience is familiar with the meaning of each symbol, little is gained by strict adherence to the standard, which often involves considerable drafting problems. For example, the wavy symbol for "punched tape" is more difficult to draw without a template.

The practical application of these symbols is illustrated in IBM's Data Processing Techniques Documentation brochure C20-8152-1, 1969, entitled "Flowcharting Techniques" (Technical Publications Department, White Plains, N.Y. 10601).

LAYOUT FORMS

The laying out of charts, graphs, and tables for reproduction is facilitated when layout forms with suitably preprinted marginal scales are available. Figure 7-16 gives an example. The margin provides for various sets of scales: (1) standard ruler, graduated to 1/16th of an inch, (2) typewriter lines, spaced at 6 to the inch, and (3) typewriter spacing at 10 letters to the inch (pica size) and 12 letters to the inch (elite size).

Prescaling has the advantage that (1) no further measurements are needed by the designer of the graphic materials, and (2) the relationships between lines and angles will be consistent with a square format. The marginal prescaling assures these effects. When light blue color is used for the scales the guidelines can be suppressed in the printing process, so as not to show up in the final product. The reader should be warned, however, that some modern printing methods will permit blue lines to show through.

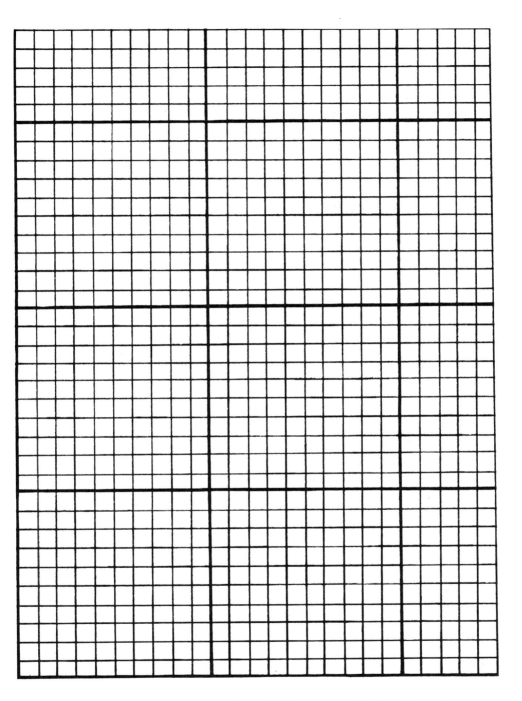

Figure 7-1: General purpose design grid. For most applications the grid will provide a convenient base on which to set up, space, and generally sketch a graph or chart.

101

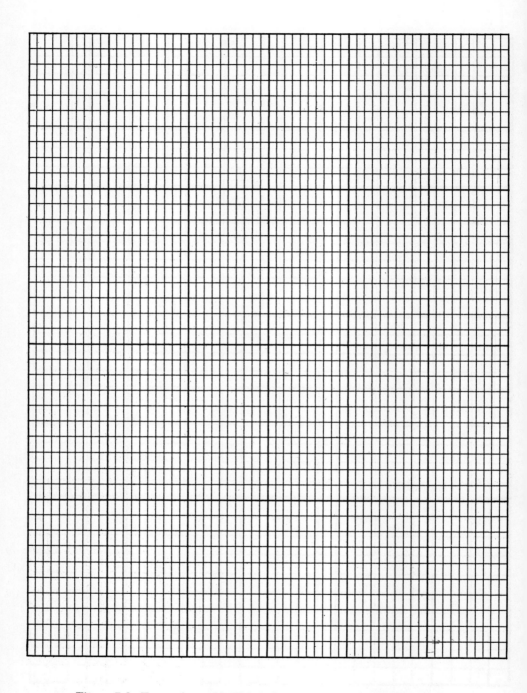

Figure 7-2: Typewriter grid. The design of graphs that allow for typing in subsequent text is facilitated. There are 12 vertical and 6 horizontal lines to the inch.

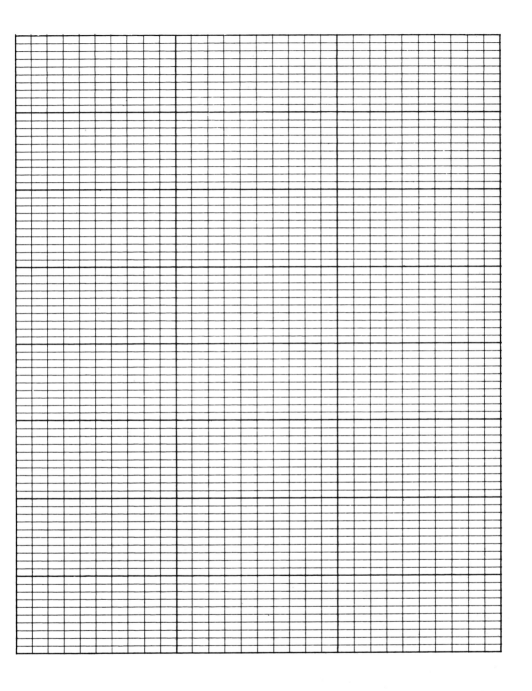

Figure 7-3: Typewriter grid designed for horizontal pages. All other detail is similar to the grid in Figure 7-2.

103

Isometric-Orthographic

Figure 7-4: Isometric grid. Three-dimensional drawings are facilitated
by this paper (which, incidentally, is usually used horizontally). Master
grid for printing through courtesy of National Blank Book Co., Holyoke,
Mass.

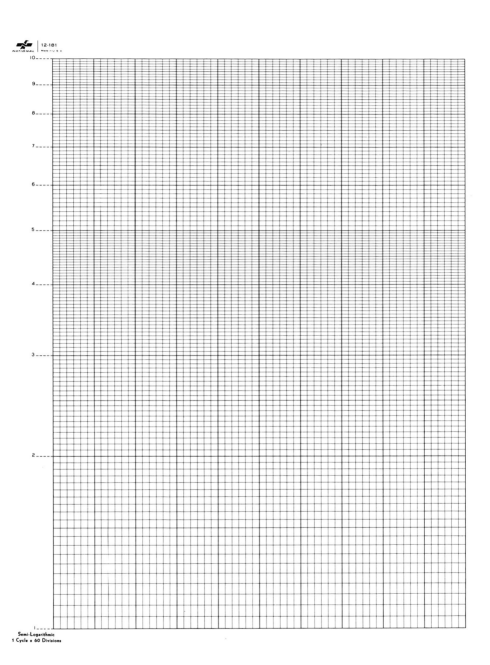

Figure 7-5: Semilog paper, one cycle, 60 division. Master copy for printing through courtesy of National Blank Book Co., Holyoke, Mass.

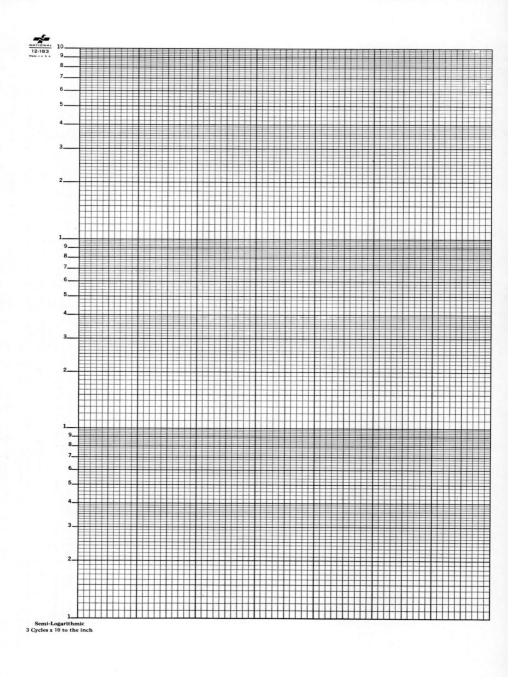

Figure 7-6: Semilog paper, two cycles. Master copy for printing through courtesy of National Blank Book Co., Holyoke, Mass.

106

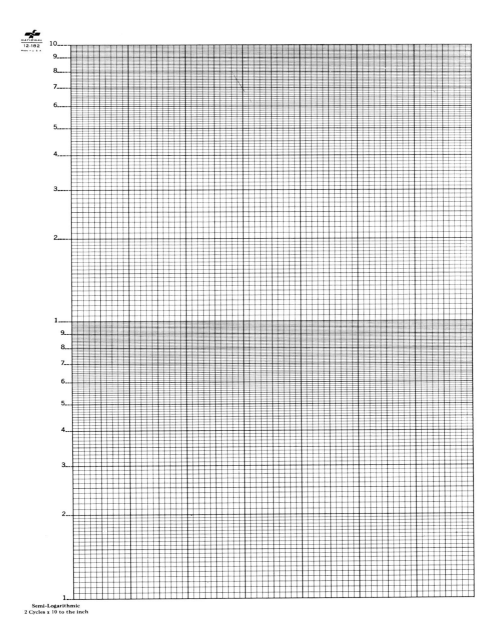

Figure 7-7: Semilog paper, three cycles. Master copy for printing through courtesy of National Blank Book Co., Holyoke, Mass.

107

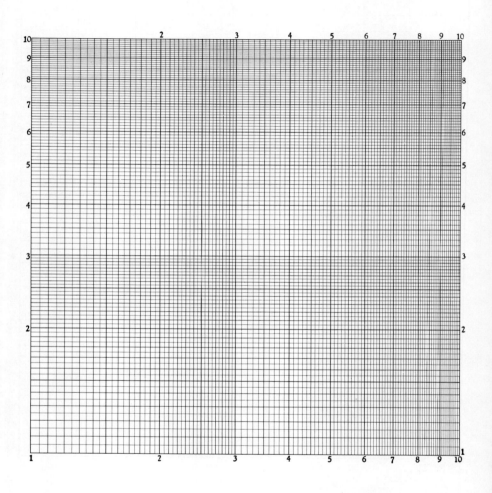

Full Logarithmic 1 x 1 Cycles

Figure 7-8: Double-log or full-log paper. Master copy for printing through courtesy of National Blank Book Co., Holyoke, Mass.

108

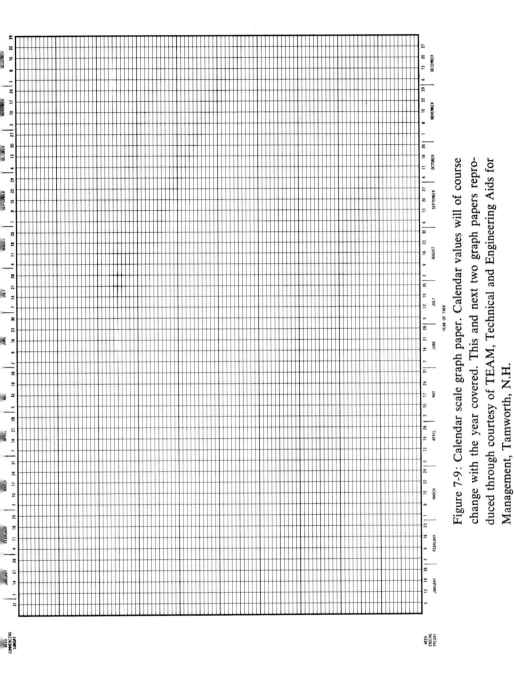

Figure 7-9: Calendar scale graph paper. Calendar values will of course change with the year covered. This and next two graph papers reproduced through courtesy of TEAM, Technical and Engineering Aids for Management, Tamworth, N.H.

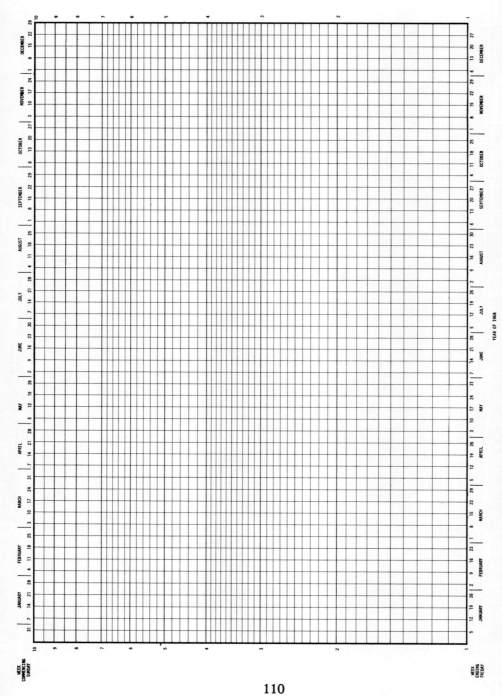

Figure 7-10: Calendar scale graph paper with logarithmic ordinate.

110

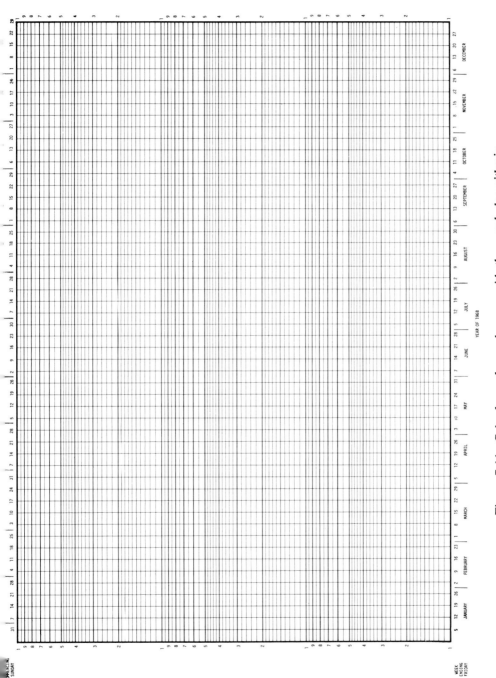

Figure 7-11: Calendar scale graph paper with three-cycle logarithmic ordinate.

111

Figure 7-12: Outline map. Statistical and other information can most readily be plotted on this outline. For example, variations in shading may indicate differences in population density. Plain numbers, location points, or volume bars may also be entered. Map through courtesy of Codex Book Company, Norwood, Mass. (stock no. 4233).

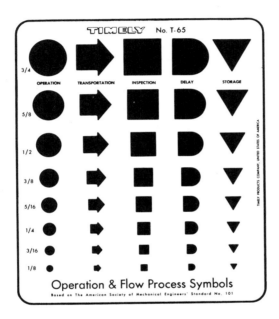

Figure 7-13: Template for operation and flow process symbols. As established by the American Society of Mechanical Engineers, the symbols refer to the following: circle = operation, arrow = transportation, box = inspection, large "D" = delay, and triangle = storage.

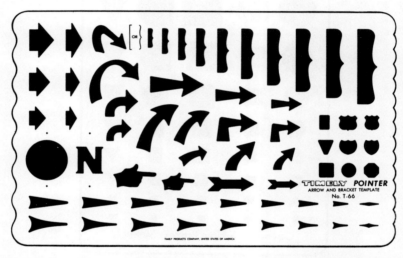

Figure 7-14: Pointer template. The cutouts provide for the drawing of arrows, brackets, indicators, and other pointers on charts.

114

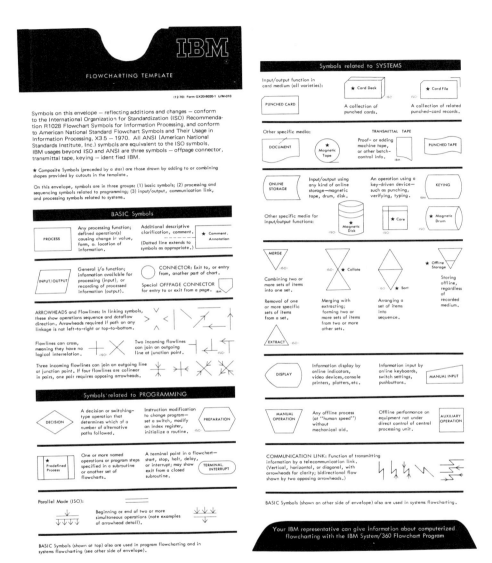

Figure 7-15: Program and flowchart symbols. These outlines are available in template form from International Business Machines Corporation. International Business Machines Corporation, "Flowcharting Techniques," Document No. C20-8152-1, 1969, White Plains, N.Y., Technical Publications Department, IBM.

115

Figure 7-16: Layout form. The marginal scaling facilitates the design of charts, graphs, and tables. Business forms can also be prepared effectively with the aid of the margins, which are scaled in typewriter spacing increments. Source: Graphic Systems, Yanceyville, N.C.

8. Tabular Presentation

"Charts" as defined in most dictionaries can be interpreted as including tabular material. Tables are a type of chart. We have previously emphasized the desirability of supplementing graphic material with more detailed tabular information, wherever this additional information is required. But regardless of whether tables support graphs or are to stand by themselves, they should be planned and designed so as to convey basic data and their analytical results in a form that highlights the conclusions and serves as a basic support for executive planning, decision-making, and control.

DEVELOPMENT OF TABULAR MATERIAL

While the need for data will vary for each organization and purpose, certain common factors go into the development of genuinely useful tables, as shown in the flow diagram of Figure 8-1. The process of generating the tables should conform with principles of clarity, completeness, and proper mechanical detail, such as are enumerated in Table 8-1. We must recognize the ever-intensifying outflow of data that is characteristic of the world of economics, business, industry, and technology today, and we must study and analyze this torrent of information to extract essential content, meaning, and significance, which is then conveyed in comprehensible form.

TYPES OF INFORMATION

Although most tables present quantitative material, information may also be conveyed effectively in verbal listings. We have already seen an example of the latter (Table 8-1), which represents a listing in tabular ar-

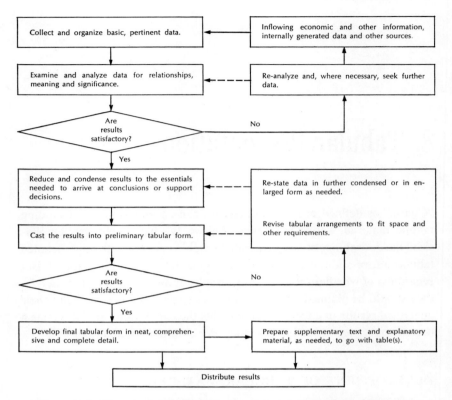

Figure 8-1: Steps in developing tabular material. A well-conceived table is the end product of a careful process involving research to gather the pertinent data, extract the meaningful information content, and present the final results in a form that highlights the conclusions or serves as a basis for managerial decision-making.

rangement (also known as a "word chart"). For numerical tabulations it is often desirable to show not just the quantitative data themselves, but also the calculation procedures that yield derived information. Finally, most numerical tabulations will contain some purely verbal portions, such as in the explanatory footnotes. The overall aim, in all instances, is to transmit information in true, though condensed, form.

DETAILS IN PREPARING TABULATIONS

In order to examine the relevant mechanics of effective tabular presentation, we will use an example that involves both basic and derived data.

TABLE 8-1

Features of Effective Tabular Material

Category	*Characteristics*
Purpose	Transmittal, in true condensed form, of the results of studies, analyses, investigations, developments, or other operations.
Clarity	Data and textual support make clear the central meaning and significance of the information.
Completeness	All the essential information, as required by technical and other factors inherent in the subject, is included. Where needed, several tables are used.
Simplicity	Nonessential detail is omitted. Simple language is used.
Balance	A balance is struck between (1) the need to assure completeness, and (2) the desire to achieve simplicity.
Layout	Material is laid out so as to effectively utilize the space available, keeping in mind the overall purpose and detailed content and organizational requirements of the table.
Style	Uniform style in general organization, identification of data, textual support, and explanatory footnotes helps the reader in focusing his attention on the information content, without distraction by minor detail.
Appeal	Attention and interest is guided toward the principal findings and results, their significance and import.
Illustration	When reference data for calculation or analysis purposes are given, it is often helpful to include a footnote providing an explicit example. Graphs should supplement tables in all instances where this aids in achieving clarity, simplicity, and completeness.
Referencing	Origins of data, source credits, and other pertinent references appear in a footnote. They help the reader evaluate the nature and validity of the data and guide him to further material.
Numbering	A table number permits ready reference, citation, and location.
Titling	Titles have all the identifying values of table numbers, while also indicating the nature of the material to follow. For referencing and immediate appeal, short titles, with more explicit subtitles, seem preferable to lengthy main titles.

119

In Figure 8-2 the illustration utilizes basic data (in rows *a* to *c*) from which a recommended "optimum lot size" for manufactured product is developed in row *d*, as part of an inventory management project. In actual use, the further data (rows *e-i*) serve to show that the proposed optimum-lot-size approach will yield substantial savings in the value (and thus carrying costs) of average inventory (row *i*), compared to the firm's existing practice. The table thus has the function not only of presenting the basic data, but also showing the cost-saving value of the optimum lot sizes recommended.

Careful attention to correct mechanical detail makes the table more professional-looking and helps the reader spot its highlights quickly. We will examine these features below.

Showing start and finish. Beginning and end should be shown clearly, particularly where there are lengthy footnotes or other explanatory items in the lower section of the table. In this way, the material is set off from the regular text. One way of accomplishing this separation on typewritten pages is to place a double line under the title and again at the end of the table. In printing, numerous approaches utilizing different sizes of print may obviate any line-separation needs.

Headings. A principal heading or title serves to identify the nature of the table. It is desirable to avoid overly long wording, resorting instead to a brief title with subheading. For example, a briefer heading than "Optimum Inventory Levels and Lot Sizes for a Manufactured Product" might have been desired. We could then have written a main title, "Recommended Lot Sizes," and a subtitle "Basic Production and Cost Data, Used to Develop Recommended Lot Sizes that Will Yield Optimum Inventory Levels." For ready identification in referencing, provide a table number with the title.

Identifying rows and columns. For effective handling of complex tabular material, it is desirable to identify column headings and row designations. In our illustration we gave row designations (*a, b, c,* etc.), which facilitated subsequent references, particularly in showing how various derived tabular quantities were calculated. In the present example there was no real need to identify column headings, but if there had been, a numbering in italicized or parenthetical 1, 2, 3, etc., might have been used.

Units of measurement. Units of measurement such as pounds, degrees, number of items, percent, or monetary values are usually best shown under each column heading or with each row designation. This practice permits showing the unit only once per column or row. A design sin committed too often is that of failing to indicate the units of measure at all, thus presenting the audience with a puzzle.

Explanatory notes. Footnotes pertaining to specific items in a table are customarily placed at the bottom of the table. Notes pertaining to all of the data should appear in parentheses near the top of the table and under the heading or subheading.

Comprehensiveness. To the extent possible, tables should include in concise form all of the pertinent information, without leaving anything dangling. Much of this effect is accomplished by assuring a certain self-explanatory quality of the material, whereby the reader observes immediately: (1) data that are given (preferably with indication of source), (2) data that are derived from the given material (preferably by showing the formula utilized), and (3) essential further explanatory information. The more self-contained the table, the more effective it will be. Few readers are likely to expend energy going back and forth between regular text and table to digest material that should have appeared in comprehensive form in the first place. Worse yet, the author who has been negligent in making his table complete will often be found similarly wanting in his regular text discussion pages.

Symbols and abbreviations. Few, indeed, are the symbols and abbreviations that are readily understood by everybody. Use the footnote section for explaining abbreviations and symbols. Better yet, see whether the full words cannot be given with the column headings and row designations themselves. Mathematical or other constants and factors also require explanation or reference to source. Finally, it seems better to explain too many symbols than to leave some unexplained.

MAKING MATERIAL FIT

A frequent annoyance in preparing tabular arrays is that the material does not fit. Resort to larger-size paper is a temptation that should be resisted. Nonstandard sizes will not fit most duplicating equipment, and the cost and time losses of special arrangements for handling off-sized materials are often critical.

By planning and replanning tabular material, changing lengthy wording to more concise form, rearranging data and removing nonessential detail, we often fit a table within regular page confines.

Table ———

Optimum Inventory Levels and Lot Sizes for a Manufactured Product

TABLE—: OPTIMUM INVENTORY LEVELS AND LOT SIZES FOR A MANUFACTURED PRODUCT

Data and Calculations	Component				
	A	B	C	D	Total
Given Data					
a. No. of components used per product	1	2	1	10	
b. No. of components used per month, based on 100 products monthly, = **a** × 100	100	200	100	1000	
c. Cost, $ per component	20	10	5	4	
Calculated Data					
d. Optimum lot size*, no. of components	100	200	200	707	
e. Ordering frequency per month = **b/d**	1	1	½	1.4	
f. Average value of regular inventory, $ = **d** × **c** × ½	1000	1000	250	1414	3664
g. 3-day safety stock = 10 per cent of **b**	10	20	10	100	
h. Average value of safety stock, $ = **g** × **c** × ½	100	100	25	200	425
i. Average value of inventory carried, $ = **f** + **h**	1100	1100	275	1614	4089

$$\text{*Optimum Lot Size} = \sqrt{\frac{\text{Monthly Requirements} \times \text{Set-up Costs}}{\text{Component Cost} \times \text{Carrying Charge} \times \frac{1}{2}}}$$

where Monthly Requirements are given in **b** and the constant ½ assumes that on the average only half the lot will be in inventory, the other half having been used in production. For this firm, inventory carrying charges are estimated at 12 per cent per year and thus 0.01 per month, based on the cost per component (in **c** above). For each of the components, set-up costs per lot are estimated at approximately $10.

Notes: Observe that a 3-day safety stock represents 10 per cent of a 30-day month, hence the use of 0.10 in **g** above.
An ordering frequency of ½ means that we order every second month, while an ordering frequency of 1.4 calls for approximately 2 orders every 3 months (see Line **e** above).

Figure 8-2: Typical tabulation that includes calculation steps. The tabulation is self-contained, in the sense that all values derived can be rechecked by the reader without resorting to separate text. Nevertheless, separate text elsewhere may elucidate and further amplify the tabular material.

A frequent design error that creates problems of fitting is the improper line-up of rows and columns. Examine the table in Figure 8-3, representing the data from Figure 8-2, but with columns and rows interchanged. Now assume that in your first try you had tabulated the data in this difficult format. No matter how much you had squeezed columns, you still would have lacked the space to accommodate the missing columns *h* and *i* (which are the easily fitting rows *h* and *i* of Figure 8-2). The experienced table designer would have spotted the problem at a glance: In Figure 8-2, the tabulation contains only four rows but must occommodate as many as nine columns. Reverse the arrangement, so that we now have nine rows on vertical paper and, *voilà*, it all fits neatly.

VARIATIONS IN USAGE

In lieu of tables, regular text or graphic presentations are often an alternative. For example, instead of the information arranged in Table 8-1,

Table ———
Optimum Inventory Levels and Lot Sizes for a Manufactured Product

TABLE—:

	Given Data			Calculated Data			
	a	b	c	d	e	f	g
Comp-onent	No. of Components Used per Product	No. of Components Used/Month Based on 100 Prod./Month $= a \times 100$	Cost per Component $	Optimum Lot Size No. of Components	Ordering Frequency per Month $= b/d$	Average Value of Regular Inventory $ $= d \times c \times \frac{1}{2}$	Three-Day Safety Stock $= 10\%$ of b
A	1	100	20	100	1	1000	10
B	2	200	10	200	1	1000	20
C	1	100	5	200	½	250	10
D	10	1000	4	707	1.4	1414	100
Total	—	—	—	—	—	3664	—

Figure 8-3: Data from the tabulation in Figure 8-2 rearranged. Note that rows have become columns, and vice versa. The vertical zig-zag line indicates that columns *h* and *i* could not be accommodated despite the use of horizontal rather than the vertical placement of the table. The lesson to be learned from these two arrangements is, when a particular setup does not permit placing all pertinent data, try interchanging rows and columns. The switch will often solve the problem, particularly when there are few rows but many column entries to be accommodated.

we could have made use of a series of paragraphs of regular text in which the various categories and characteristics of effective tabular material are discussed. As a matter of fact, some authors prefer this more lengthy and tedious path of communication to the concise listing that was attained by the tabular form. No hard and fast rules apply, of course, as to which alternative is preferable or applicable. In some instances both tabular and textual presentation will be utilized.

The relation between tabular and text alternatives is parallel to that beween diagrams and text. We need only examine Figure 8-1 to observe that the flow of information shown therein could have been alternatively given in text form. Moreover, with some ingenuity the material can be reshaped into a series of steps that can be listed as part of a table.

Special devices are appropriate on certain occasions. Thus in this chapter we have resorted to the unusual procedure of putting illustrative tables within the borders of a rectangle that we then call a "figure." The rationale for this procedure is to distinguish these illustrative samples of tabular form, such as in Figures 8-2 and 8-3, from regular tables that form part of the principal discussion, such as Table 8-1.

Further underscoring the interchangeability of usage of media of communication is given in Figure 8-4. This represents a functional table of contents, emphasizing how a book is used. The sequence in which chapter

ORGANIZATION AND USE OF THE HANDBOOK

THE FOLLOWING TABLE IDENTIFIES APPLICABLE SECTIONS WITHIN THE HANDBOOK CORRESPONDING TO THE MAJOR RELIABILITY ASSURANCE FUNCTIONS TO BE PERFORMED.

TO PERFORM THESE RELIABILITY FUNCTIONS →   USE THESE SECTIONS OF THE HANDBOOK →

	1	2	3	4	5	6	7	8	9	10
DEFINE REQUIREMENTS		●								
ESTIMATE FEASIBILITY		●								●
ALLOCATE RELIABILITY		●								●
PREPARE A TDP				●						●
PREPARE A SPECIFICATION				●						
PREPARE AN RFP				●						
ESTIMATE TIME AND COST								●		
PREPARE CONTRACT TASK STATEMENT				●						
FORMULATE A DESIGN		●			●					●
REVIEW A DESIGN					●	●				
EVALUATE DESIGN PROBLEMS					●	●				
EVALUATE A PRODUCT OR PROCESS			●			●		●		
DESIGN AN ACCEPTANCE TEST							●			
PLAN A RELIABILITY PROGRAM			●							
MONITOR A RELIABILITY PROGRAM			●					●		
USE A RELIABILITY "FEEDBACK" LOOP								●		
MAKE A FAILURE ANALYSIS								●		
MAKE A FIELD EVALUATION								●		
CONDUCT A TRAINING COURSE	●	●	●	●	●	●	●	●	●	●
MANAGE A RELIABILITY PROGRAM		●								

Figure 8-4: A table of contents in graphic form. A combination of tabular arrangement and flow chart directions serve to make this material attractive and useful. From "Proceedings of the Naval Material Support Establishment Systems Effectiveness Conference," April 1965, U.S. Naval Applied Science Laboratory, Systems Effectiveness Branch, Brooklyn, N.Y.

materials or sections of the book occur is also shown, but this information is made subordinate to a consideration of the pertinent functions and the usage the reader has in mind. Note that the book is divided into 10 sections that are the equivalent of "chapters" in the ordinary context. This is a minor divergence from regular practice. The important factor is that the author chose to prepare a chart relating various functions—reliability-assurance functions, in this instance—to the places in the book where these topics are covered. Here again, a functional table of contents need not supplant a regular table. Both may be given.

The resourceful communicator will realize that he has at his disposal a variety of methods and techniques for presenting materials. There are graphic and tabular vehicles; within these two categories there are a multitude of types of application. In choosing among the available forms and possible uses, one must maintain an open mind and a high degree of flexibility. The manner in which graphs, tables, and text are used in various combinations, to what extent there should be overlap and redundancy, and related matters will hinge on two factors: (1) the weight, detail and complexity of the material to be conveyed, and (2) the anticipated abilities of the intended audience to follow, grasp, and meaningfully understand the information presented. Usually, when writing to a well-informed group on a noncomplex subject, the simplest approach, using a minimum of graphs, tables or text, is called for. To do anything else will risk the reader's boredom by the low density of thought per page. On the other hand, when the nature of subject matter is such as to be difficult in relation to the reader's background, then the communicator may well decide on a massive approach. For the sake of attaining and holding attention, he will rely on many well-designed charts and graphs. To the extent really needed, he will supply additional or more detailed data in tabular form, even though this will to some extent duplicate the graphs. Finally, he will write his text as clearly and fully as possible, even though in many instances the text will be redundant with graphs and tables. Much of the information is thus conveyed in three parallel forms, but the space consumed and the effort expended are justified. Complex material may indeed have to be viewed from many aspects and angles until it can be fully grasped.

MISCELLANEOUS FURTHER USES

In the following are present a number of additional illustrative examples of tabular form.

TABLE
Dynamics of Pricing for Inventory Control

1. *Planning Process*

Long-range forecasts are obtained, based on market research, on estimates by sales executives and salesmen and on analysis of general market and business trends. Based on these forecasts, expected quantity volumes demanded, prices considered maintainable, and risk factors in inventories for future sale are developed. Long-range forecasts usually cover a year or a 3-month or 6-month season and are developed by various product groups, product lines, and, where required, individual items.

2. *Short-Range Adjustments*

As the season gets under way, the progress of sales is watched closely. Comparisons with past seasonal patterns of buildup of sales and a review of unsold capacity leads to short-range forecasts, subject to weekly review and revision.

3. *Decisions*

Based on the long-range plans and long-range forecasts, adjusted for current developments, daily decisions are made on volume of production, pricing (acceptance and refusal prices), and buildup or depletion of inventories.

4. *Review of Results*

The effect of management's production, pricing, and inventory decisions is watched closely. The success in keeping plant capacity sold, maintaining prices, and avoiding risky inventory positions is evaluated. Revisions are based on these weekly reviews—weighing forecasts, decisions, and results in their overall effect. Future decisions in pricing, for example, are based on the effect of prior pricing actions. Also, when required on the basis of the evidence accumulated, short-range expectations and long-range forecasts are revised.

5. *Overall Results*

As established, the system utilizes new information (market research, sales estimates, sales progress, and related data) for its planning processes. In addition, it takes advantage of hindsight (the experience gathered from the effect of actual decisions) for guiding foresight (revising forecasts, adjusting decisions). Management thus has a *dynamic procedure* to suit the dynamic factors of the market, blended with an *adaptive process* to revise decisions in the light of their effects.

Figure 8-5: Tabular form of flow chart information. If the five paragraphs representing successive steps of a system are put in blocks and connected with arrows, from step 1 to 5, a flow diagram results. Both approaches—tabular and charting—are effective. Note, however, that it is difficult to show feedback relationships in the tabular version. This and Figures 8-6 through 8-8 are from N. L. Enrick, *Inventory Management,* Scranton, Pa.: © Chandler Publishing Co., a division of Intext, 1968.

Tabular Flow Chart. A flow chart can be given in tabular form, as shown in Figure 8-5. Had we placed the five successive steps of this table in separate blocks, connecting the blocks by arrows, we would have had a flow chart. In this example, the flows are relatively straightforward. When intricate sequences must be traced, a tabular approach will usually be inadequate.

TABULAR PRESENTATION

List. Illustrated in Figure 8-6, a listing tabulation represents a convenient schedule for presenting individual items within meaningful groupings and sequences. The example catalogs two major types of cost factors, each of which is subdivided according to an inventory of individual contributing items. In the lower section of the table, a schedule of other, related factors is given. Such terms as record, roll, and register are occasionally used to denote a list.

Tabular glossary. When a glossary is relatively brief, it may be con-venient to include it within the confines of a table, such as in Figure 8-7. This approach has a distinct advantage over simple discussion of each term within regular text. The table specifically asks for a listing of the symbol for each term, the units of measurement in which it is expressed, and a definition in which the term is explained and example of its use given. Any gaps in the table will tell the author that he has made an omis-sion. The reader benefits because he can refer to the symbols, units, and definitions more readily than if he had to work through the text.

Check List. Figure 8-8 is an example of a check list in which various outcomes (error situations in this example) are related to a set of possible situations (cases 1 to 10 in the example). The idea of this check list is that the reader would review the schedule of cases and related situations against his own possible decisions, with a view to reconsidering any decisions that may appear undesirable. When check lists are prepared with check boxes, such as in Figure 8-9, we refer to them as forms or figures rather than tables.

TERMS DEFINING THE STRUCTURAL PARTS OF A TABLE

Although there is no real need for formal definitions in the preparation of

TABLE 7

Factors to Be Considered in a Large-Scale Simulation

Basic Cost Factors

1. Inventory carrying, including storage, insurance, interest on money tied up in inventory, obsolescence, handling, spoilage.
2. Costs of order placing, including shipping charges for less than optimum quantities, clerical and accounting, special trips, and related costs.
3. Loss in revenue from unavailability of merchandise.

Implied Cost Factors

1. Loss in goodwill from unavailability of merchandise.
2. Risks of excess stock left over at end of season.
3. Effect on employee morale of frequent stockouts, such as
   a. Interruptions in production.
   b. Use of substitute materials that give production and quality problems.
   c. Generally disorderly way of doing business.
   d. Emergency meetings, long-distance calls, excessive bothering of suppliers.
   e. Waiving of quality requirements when inferior supplies arrive but deadlines demand their use anyway.

Allowance for Various Managerial Remedies or Special Situations

1. Willingness of some customers to wait when no stock is on hand.
2. Ability to sell customers on other merchandise that is in stock.
3. Possibility of special emergency shipments in case of stockouts.

Figure 8-6: Listing tabulations. This table lists or documents a number of items. Here the items happen to be cost and related factors considered in a computer simulation of inventory control. The availability of such a list facilitates review and analysis of the factors considered in the actual systems work pertaining to this project.

tables, it may nevertheless be of interest to examine the terms in Figure 8-10, identifying the various structural parts involved. The left-hand side of the illustration provides an example of a table in which various segments have been labled. The tabular layout is repeated on the right-hand side, but this time we show the terms applicable to each entry in the sample table to the left. The nomenclature is based on the recommendations of the Bureau of the Census.*

* U.S. Department of Commerce, Bureau of the Census, *Bureau of the Census Manual of Tabular Presentation. (a special report of the Bureau of the Census)* Washington, D.C.: U.S. Government Printing Office, 1949, 265 pages.

A particular feature of the tabular arrangement is that the totals for various columns and lines appear before the detailed entries. It should be realized that this procedure need not be followed. In fact, many times it will be found preferable to show detail first and totals later. Other modifications in the arrangement of columns, lines, blocks, areas, and fields of tables may be made based on the purposes and needs at hand. Nevertheless, the arrangement in Figure 8-10 is a good reference point for tabular structuring.

We may amplify the definitions by looking at the functions and purpose of the various parts of a table.

Title: A brief indication of the nature and scope of the data presented.

Headnote: Information supplementing the title. When information pertains to a segment of the table only, utilize a footnote for the explanatory material.

Stub: A listing of line or row captions or other descriptions, together with needed classifying and qualifying heads and subheads. The stubhead or box is the column head of caption of the stub. It describes the stub listing as a whole.

Center head: A classifying, descriptive, or qualifying statement applying to all subheads and line captions below it.

Line caption: The description of the data appearing on the line.

Block: A group of related line captions with their heads and subheads.

Boxheads: The area that contains the individual column heads or captions describing the contents of each vertical row or column.

Spanner head or caption: A classifying, descriptive, or qualifying caption spreading across and above one or more column heads.

Panel: A segment of the boxhead, consisting of a group of related column heads.

Field: The portion of the table that extends from the lower end of the boxhead to the end of the table and to the right of the stub (assuming that, as usual, the stub is on the left-hand margin of the tabulation).

Cell: The intersection of any line and column.

Line: A horizontal row of cells.

Column: A vertical row of cells.

Only the essential terms have been presented. The Bureau of the Census manual presents several dozens of pages on this topic. The list here is confined to those terms needed for most practical purposes.

TABLE

Factors for Determining Optimum Lot Size

Term Used	Symbol	Unit	Explanation
Lot size	Q	$	Quantity in a purchase or production order. An order calling for 10 bearings at $20 per bearing has a lot size, $Q = 10 \times \$20 = \$200$. Under a steady consumption or sale of these bearings, the average annual inventory will therefore be $Q/2 = \$200/2 = \$100 = 5$ bearings.
Inventory carrying	I	decimal fraction	Cost, as a decimal fraction of inventory value, of carrying inventory in stock, including interest, obsolescence, depreciation, spoilage, storage, taxes, and insurance. If it costs 10 percent annually to store a bearing, then $I = 0.1$. Annual inventory costs in dollars are found from $IQ/2$. For the bearings, $IQ/2 = 0.10 \times \$200/2 = \$10$.
Frequency	F	number	Frequency per year that an order is issued. If bearings are ordered monthly, then $F = 12$; if yearly, then $F = 1$.
Yearly usage	Y	$	Yearly consumption of supplies or yearly cost of goods sold. For example, annual use or sale of bearings may be 120 bearings at a cost of $20 each, so that $Y = 120 \times \$20 = \$2400$. Note also that $Y = FQ$, so that $Q = Y/F$.
Procurement cost	P	$	Costs incurred each time a production order or purchase order is issued. Often issuance of an order involves merely the cost of paperwork, bookkeeping, and followup; but a production order involving changes in machinery setups may also involve high labor and materials costs. Thus, while an individual purchase order may cost from $2 to $4, a production order may incur additional costs of from $10 to $500, depending on the item involved. For a bearing, if $P = \$4$, then the ordering cost per annum will be $PF = \$4 \times 12 = \$48$. Also, since $PF = PY/Q$, $\$4 \times \$2400/\$200 = \$48$.
Optimum lot size	Q_0	$	Optimum lot size balances the costs of I and P, using $Q_0 = \sqrt{2PY/I} = \sqrt{2 \times \$4 \times \$2400/.01} = \$440$ rounded. This $440 represents 22 bearings at $20 each. Moreover, the optimal frequency, $F = Y/Q_0 = \$2400/\$440 = 5$ to 6. As a practical rule, we would order every two months.

Figure 8-7: Tabular glossary. When a glossary is relatively small, the pertinent terms, symbols, definitions, or explanations can be provided within the framework of a table.

TABLE

Errors in Rule-of-Thumb Inventory Decisions

Case Number	Actual Situation Prevailing	Decision Made	Error Involved in Decision
1	Low unit cost per item	Order small quantities	Frequent ordering of items will raise clerical and bookkeeping costs and may skyrocket transportation (shipping-in) expenses.
2	High unit cost per item	Order large quantities	Needlessly high costs of carrying large inventories for long periods, risk of obsolescence, deterioration, and spoilage during storage.
3	Short delivery-time items	Place orders too early	Storage space is tied up, inventories are inflated, and spoilage before use or sale may occur.
4	Long delivery-time items	Place orders too late	Risk of stockout, affecting sales or production or both.
5	High setup costs	Schedule short runs	Since setups are part of the total costs of production runs, unduly high product costs may result.
6	Low setup costs	Schedule long runs	Buildup of stock may create handling and storage problems and needless charges of carrying inventory.
7	Declining market demand	Build up inventory	Surplus goods that may have to be sold at distress prices.
8	Expanding market	Reduce inventory	Lost sales, customers, and markets because of inability to deliver goods.
9	Seasonal products	Keep inventory constant	Maximal overtime production during season will usually not be able to keep up with peak weeks or months of season, and there will be no seasonally built-up inventory to draw on.
10	Fashion items	Maintain large inventories	High risk of obsolescence. (The term "fashion" is extended to also cover items likely to be affected by fluctuating states of new technologies or other unforeseeable, rapid market changes for particular models or styles.)

Figure 8-8: Tabulation serving as a check list. Ten possible situations, or "cases," are presented, together with their implications. It is relatively easy to check each case. An alternative approach is to present the information in the form of a lengthy paragraph.

131

TABLE 8-2

Check List for Tabular Design

Item No.	Item	Check
1a	**Title**	
1	Preceded by table number.	_____
1b	Placed atop.	_____
1c	Brief but adequate for identification.	_____
1d	Subtitle explains further.	_____
1e	Reference to further explanations and qualifications, in a footnote, indicated by asterisk or similar device keyed to title or subtitle.	_____
2	**Data**	
2a	Column and line headings identifying nature of data.	_____
2b	Units of measurement (lb., in., etc.) indicated.	_____
2c	Headings and subheadings succinct in identifying data.	_____
3	**Footnotes**	
3a	Symbols and abbreviations explained.	_____
3b	Supplementary notes and explanations and qualifying statements provided.	_____
4	**Spacing**	
4a	Generally, one table per page (except really brief and usually unnumbered tabulations or printed material).	_____
4b	Regular-size paper, unless larger size is really needed.	_____
4c	Double-size page where regular size is inadequate.	_____
4d	Margin for binding and enough room on other margins to enter page number, etc., as required by mechanical and appearance considerations.	_____
5	**General**	
5a	All essential data given to support the meaning and message of the tabulated data, in readily understandable form, for the intended audience.	_____
5b	Self-explanatory qualities. Titles, headings, and related text should be clear, concise, and sufficiently informative to allow the table to stand by itself—even though other text may further elucidate.	_____
5c	Uncrowded and uncluttered appearance. (Is it necessary to break a single table into two less cramped tables? Can we simplify it?)	_____
5d	Enough information per table. Avoid formal tables with diminutive amounts of information. Possibly several tables should be combined into one.	_____
5e	Proper proportioning. If table is crowded in one area and white space abounds in other segments, then the arrangement may benefit from review and revision. Lettering, too, should be properly proportioned in relation to relative significance. Avoid dominant headings that detract from the overall impact. Undersize lettering is equally undesirable.	_____
5f	Design and redesign—until the table is satisfactory.	_____

How would YOU rate the service of
THE DEALER WHO SOLD YOU YOUR CAR?

We would appreciate your ratings on the dealer <u>who sold you your car</u>, even if you have not used his service facilities for some time.

Please ☑ *check answers*

	Good	Fair	Poor
Clean and orderly car receiving department?	☐	☐	☐
Do they wait on you promptly?	☐	☐	☐
Correct "diagnosis" of trouble? *(good at finding out what's wrong)*	☐	☐	☐
How about quality of work?	☐	☐	☐
Accurate estimates of cost? *—don't usually get an estimate—* ☐	☐	☐	☐
Phoning you about unexpected repairs needed?	☐	☐	☐
How about attention to minor details?	☐	☐	☐
Work completed when promised?	☐	☐	☐
Quick service on lubrication & minor jobs?	☐	☐	☐
Work completed satisfactorily first time? *(so you don't have to return for the same job)*	☐	☐	☐
Parts availability in dealer's stock?	☐	☐	☐
Are grease and dirt cleaned off steering wheel, windshield, etc?	☐	☐	☐
Do they seem to appreciate your patronage?	☐	☐	☐
Are they courteous and understanding?	☐	☐	☐

Are the service department prices fair?   ☐ YES   ☐ NO

On the Whole

—how would you rate the service of the dealer who sold you your car?

☐ GOOD   ☐ FAIR   ☐ POOR

Figure 8-9: Check list in questionnaire form. This appearance of a check list usually falls under the category of a "figure" or "diagram," although it may also be referred to as a table. From N. L. Enrick, *Cases in Management Statistics,* New York: Holt, Rinehart & Winston, 1962, p. 101.

133

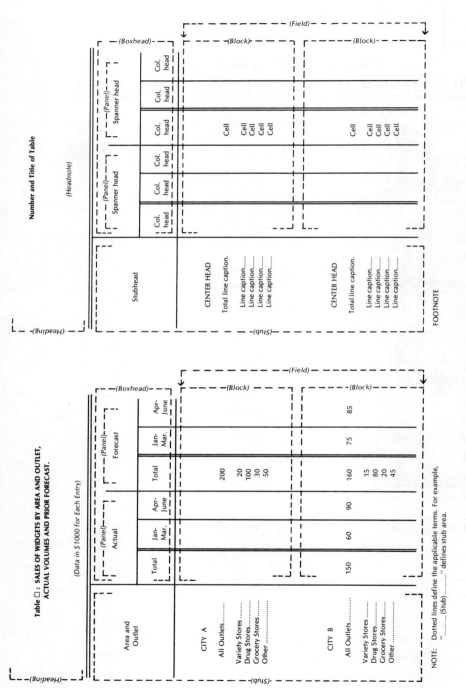

Figure 8-10. Terms used to define structual parts and entries of a table. On the left is an illustrative example. On the right are the terms applicable to each entry.

SUMMARY

Quantitative data from studies, investigations, and research are shown comprehensively in tabular form. But unless it is well thought out, material that should be informative may only be confusing. Whether an executive is to be sold on significant findings, or a recommended program is to be supported by facts, or a customer convinced of the superior characteristics of your product, data must be arranged so as to highlight the truth informatively and persuasively.

9. Decision-Structure Tables

Tables that are well designed and executed contain a wealth of information in comprehensive form, thus facilitating review, study, and detailed analysis. From the conclusions reached on the basis of this data, the user of tabular material is in a better position to formulate plans and arrive at decisions. We have emphasized the decision-furthering end result of tables, since an organization may well be likened to a beehive of such activities, where decision-makers must cope with numerous choice-and-selection problems in accounting, finance, sales, and production work.

DECISION DOCUMENTATION

Ordinary tabular material helps the decision-maker, it brings the data together, and hopefully it includes useful ratios and other analytical results. Yet each decision must be made not just from the facts and figures themselves, but also in light of prior actions that have built up an intricate structure of minor and major routines. In turn, these procedures are based on a multiplicity of detail and well thought-out logical sequences of considerations. Unless decision-structure information has been documented, however, it is unlikely that one can readily find a comprehensive analysis and related instructional or historical material on this subject. Instead, such substantiation as may exist is based on past usage and experience, people's recollections, vague memories, scattered records, a variety of files, and other bits-and-pieces and difficult-access sources.

In order to promote consistent logic and usage in this multitude of decision processes, the Computer Department of General Electric Company has developed a systems documentation technique known as "Decision Structuring" or "Tabular Decision Logic," and incorporated in a systems-oriented computer language known as "Tabsol," which has found increasing adoption in business and industry. Core of the approach of systems documentation is a specially designed decision-logic or decision-structure table, which presents the decision parameters of a problem in condensed and readily reviewable and usable form. Economy, speed, and accuracy of managerial performance can often be enhanced markedly through the use of this technique.

PROGRAMMED DECISIONS

The pervasive usefulness of decision structuring springs from the fact that by far the most predominant types of choice-and-selection problems handled in daily business operations are routine in nature. We may call them "programmed," in the sense that a decision structure exists for this problem, based on prior similar instances. Programmed decisions may be distinguished from novel, nonprogrammed ones in terms of the contrasts of Figure 9-1. Note, however, that at least some aspects of a novel decision process involve features of a routine nature, where prior instances of action and related experience can be of value.

Tabular decision logic serves to document for ready reference and use the generally preferred considerations affecting the parameters of a programmed decision.*

The traditional programmed decision approach, as illustrated by the flow chart in Figure 9-2, must of necessity rely on recollection of experience, handed-down instructions, word-of-mouth, and reference files and data that may be scattered in various locations. Such an approach is less precise and hardly as unequivocal as a documented decision structure would permit. The latter is the less cumbersome and less costly approach, which, moreover, should leave less room for misinterpretation and error.

AN ILLUSTRATIVE EXAMPLE

A typical, though simplified and abbreviated, decision structure chart appears in Figure 9-3. It refers to a quality control problem. An incoming shipment of components, after being inspected at the plant, has been found

---

* See T. F. Kavanagh, "Decision Structure Tables—A Technique for Business Decision Making," *Journal of Industrial Engineering*, vol. 14, no. 5 (September-October 1963), pp. 249-258.

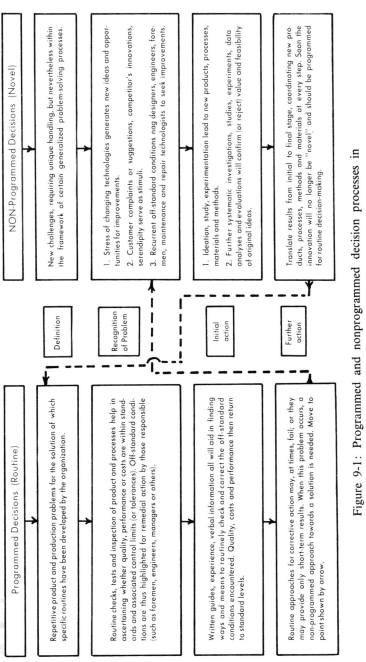

Figure 9-1: Programmed and nonprogrammed decision processes in production. Note the points of interrelation: unusual problems encountered under routine operations may lead to nonprogrammed decisions. Also, once a nonprogrammed decision problem loses its novelty it begins to fall within the domain of programmed decisions, involving routine and repetitive procedures in the future. It is recognized that in practice there need not be a clear-cut distinction between the two types of decisions shown. Programmed decisions will often entail some novel problems or aspects to be dealt with, and likewise some types of routine structures will always apply to certain portions of a nonprogrammed decision process.

below specified standards and tolerances. The decision question is now: what shall we do with the lot? A multiplicity of alternatives and action configurations must be considered, and the most predominant set of 14 such items tabulated. As an example, let us take configuration 8 and read what the table has to say: "*If* the defects are serious, *and* they cannot be repaired, *and* the plant does not need the components urgently . . . *then* return the shipment to the vendor." Here *"if"* refers to the problem condition, while *"then"* indicates the decision category. Observe also that the decision rules correspond to the problem configurations. Thus configuration 8 gives rise to rule 8. In many decision-table applications, as a

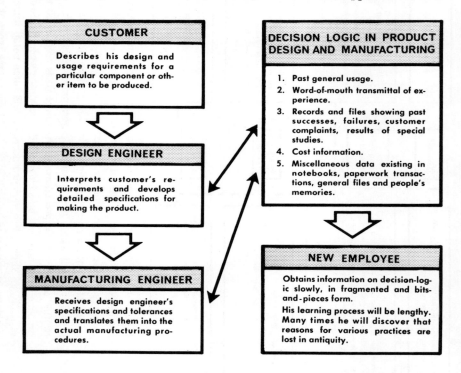

Figure 9-2: Organization and transmittal of decision-logic, traditional approach. This illustration comes from a product-design application. While many repetitive types of decision processes may have essentially sound and logical foundations, this information tends to be submerged in multitudes of records, files, and lost detail. Much of the logic may have been carried in people's heads and thus become obscured in the course of time. Systematic documentation of decision-logic avoids this problem, thus sharpening the efficiency of actual usage and speeding the learning process of new employees.

PROBLEM CONDITION	PROBLEM CONFIGURATIONS (NO. 1 TO 14)													
	1	2	3	4	5	6	7	8	9	10	11	12	13	14
a. Are defects serious?	O	O	O	O	O	O	O	O	●	●	●	●	●	●
b. Can they be repaired?	O	O	O	O	O	O	●	●	O	O	O	O	●	●
c. Is repair labor available?	O	O	●	●	O	O			O	O	●	●		
d. Does vendor agree to pay for repairs?	O	O			●	●			●	●				
e. Does plant need the product urgently?	O	O	O	●	O	●	O	●	O	●	O	●	O	●
DECISION CATEGORIES	DECISION RULES													
f. Return shipment.			✓	✓		✓	✓	✓		✓		✓		✓
g. Use product but seek allowance from vendor.											✓		✓	
h. Repair product and use it.	✓	✓			✓				✓					
i. Seek urgent replacement elsewhere.			✓			✓								
j. Go to further tables for more instructions.			I			II	I		II					

Code: O = "yes," ● = "no," ✓ = "make this decision."

Figure 9-3: Decision-structure chart. Answers "yes" or "no" to the questions posed under "Problem Condition" and identifies 1 to 14 possible problem configurations. Next, the action to be taken is specified under "Decision Rules," as drawn from five categories. In instances where supplementary guidelines are needed, reference to further tables (line *j*) is made.

matter of fact, the concept of "configuration" is merged with the rule, so that the heading "rule 8" covers both the configuration and the instruction.

With traditional procedures and without a structured table to refer to, the ultimate decision might well have been identical to the one just shown. But there is an advantage to written documentation, particularly when new employees are at work or the person normally in charge of these particular types of decisions is absent or where conflicts arise as to what has actually been past procedure, is standard practice, is "best," or is "preferred policy." Moreover, the comprehensive tabular form may save time. In-

CATEGORY OF OPERATING REQUIREMENTS	PERFORMANCE REQUIREMENT COMBINATIONS (1 TO 6)					
	1	2	3	4	5	6
SPEED OF DRIVE SHAFT, rpm	25		50		100	
OPERATING LOAD (HEAVY OR LIGHT)	HEAVY	LIGHT	HEAVY	LIGHT	HEAVY	LIGHT
DECISION CATEGORIES	DECISION RULES					
USE DRIVE GEAR NO.	101	102	106	102	108	106
ASSEMBLE AS SHOWN IN DRAWING NO.	1001	1001	1002	1002	1003	1004
NOTE FURTHER INSTRUCTIONS BY REFERRING TO TABLE:	A-115	A-115	A-116	A-116	A-117	B-108

Figure 9-4: Decision logic for product design. For example, combination 5 says: "IF drive shaft runs at 100 rpm and operating load is heavy, THEN drive gear 108 is needed. Assemble as per Drawing 1003 and additional instructions of Table A-117." Reference to other tables is common. In fact, usually a network of references and cross-references among the tables provides the complete design information. For each product, with its various requirements, we must follow the decision-logic to arrive at the specific set of design specifications that is appropriate. The tables provide a guide through this maze of requirements.

stead of having to scurry around to find the responsible decision-makers —under the traditional approach a meeting of several supervisors, managers, or executives may be required—the action to be taken is relatively automatic. There is not even a need to search through scattered records, files, and paperwork to hunt up the needed instructions and related information.

FURTHER ILLUSTRATIONS

The wide range of applicability of tabular logic is illustrated with the aid of two further examples. Figure 9-4 shows the decision structure approach applied to a product design process. Some firms have developed elaborate networks of design-logic tables, so that a large proportion of all design operations can be programmed. Even for novel designs involving new equipments, machinery, components, or other items, there will be certain

OPERATING CONDITION (FOR ROLLER BEARINGS)	OPERATING CONFIGURATIONS (NOS. 1 TO 12)											
	1	2	3	4	5	6	7	8	9	10	11	12
SPEED, rpm	UNDER 500				500 TO 1000				OVER 1000			
TEMPERATURE, °F	32-150		150-200		32-150		150-200		32-150		150-200	
DUST CONCENTRATION LEVEL	Lo	Hi	Lo	Hi	Lo	Hi	Lo	Hi	Lo	Hi	Lo	Hi
DECISION CATEGORIES	DECISION RULES											
GREASE TO USE, GRADE NO.	2	2	2	2	1	1	2	2	1	1	2	2
SHOULD EXPEL-TYPE FITTING BE APPLIED?	No	Yes	No	Yes	No	Yes	No	Yes	No	Yes	No	Yes
DRAIN-WASH-FLUSH INTERVAL, HOURS	70	50	50	40	35	25	30	20	25	15	20	10

Figure 9-5: Decision chart for bearing maintenance. For configuration 3, for example, the chart says: "IF roller bearings are operated at below 500 rpm and temperatures are between 150 to 200 degrees Fahrenheit, THEN use a Grade 2 grease, do not apply the special expel type fitting and drain, wash, and flush bearing after each 50 hours of operation." Confiurations 1 to 12 thus give rise to 12 rules.

substructures that are built from routine applications. The substructures then yield to decision logic that has been established in advance. The ultimate has been achieved in a number of instances, where many design functions are on an automated, computerized basis—with the data processing built on decision structure tables. Design engineers' time is thus saved and diverted to those problems requiring a truly creative and novel attack.

Maintenance work is another area in which a multitude of routine decisions occur, thus calling for decision structure tables. Figure 9-5 provides an example, establishing rules for the maintenance of bearings under a variety of operating conditions.

Accounting, finance, sales, marketing, and distribution are other areas where decision tables can be of great value. The approach is also of considerable value in research work in the social, biological, and physical sciences, where factor configurations need to be studied in relation to outcomes. Here tabular logic permits ready comparison of the data and a grasp of the internal structure of the system under study. In the field of applied mathematics and statistics, we have used decision tables to show a variety of different sequences of calculations, depending on various problems that needed analysis. The user then merely locates the column pertaining to his problem and follows the calculation steps prescribed by the decision rules for the problem condition.

CLARITY, CONCISENESS, AND MEANING OF DECISION CHARTS

Having examined the uses of decision charts, let us review their value in the light of the insights gained.

Clarity is accomplished by placing all of the decision parameters side by side and identifying that *if* certain configurations do prevail, *then* certain decisions should follow. The written approach helps in (a) obtaining initial understanding of the decision structure on the part of those in the firm who must approve each decision rule that becomes part of the table, and (b) in minimizing dangers of error and confusion in actual use.

Conciseness accrues because the tabular arrangement accommodates information that would require much more space in ordinary text form. Completeness is often attained as part of the concise presentation, since a quick overview of the columns and rows will spotlight any omissions, gaps, or missing items.

Meaningful relationships are demonstrated in the tabular structure by aligning problem configurations and decision rules side by side. A comparison among alternatives will reveal cause-and-effect interplays. Actions reflecting similar or related problem factors are highlighted by means of side by side comparisons, thus displaying the effects of various factors on dependent rules and situations. Moreover, when a novel decision problem is encountered, a review of the clear, concise, and meaningfully presented relationships among the factors entailed in prior decisions will soon point a way to consistent new decisions.

GUIDES VERSUS RULES

The question is sometimes raised: are decision tables guides or rules? In answer, it should be pointed out that the structuring shown merely reflects the consensus of responsible decision-makers regarding the kinds of action that logically flow from certain operational problem configurations. In that sense, the table is a prescription, an instruction, even "law," as opposed to anarchy or vagueness in decision processes. When it happens that a manager, supervisor, or even an operator feels that a certain rule might not properly apply in a given instance, or when it appears that the passage of time or changes in general conditions have outdated a decision rule, then it is obviously time to appeal for a new ruling. New decision rules are called for, rules that cover changed conditions, special circumstances, or unforeseen factors. Development of these new rules is then governed by the same procedures that went into the original set. There is obviously a loss of time as decisions are structured, but the cost of the loss is usually worthwhile in light of the relatively precise programming of future decisions that is achieved.

SUMMARY

Considerable time and effort can be saved, particularly at the level of middle management, when routine or programmable decisions are incorporated in a systematic structure—the decision table. The approach considers various possible problem configurations and then identifies the decision rules applicable to each alternative situation. With the ever-increasing multiplicity and complexity of decision problems in an organization, decision tables are an effective aid to economy, speed, and accuracy in the choice-and-selcetion process of managerial responsibilities.

10. Projection of Graphs and Tables

The charts, graphs, and tables discussed thus far have been considered primarily from a view of use in reports and similar applications, where reading is at a relatively close distance. For projection on a screen, however, some additional factors must be considered. These relate to the need to assure proper dimensioning of detail, particularly symbols and lettering, for legibility at maximum distance.

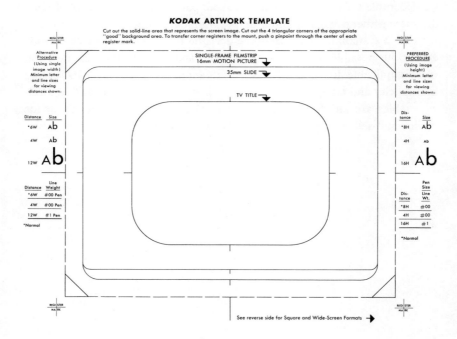

Figure 10-1: Art work template for lantern slide, 16 millimeter motion picture, and television screen. Template courtesy of Motion Picture and Education Markets Division of Eastman Kodak Company, Rochester, N.Y.

ARTWORK TEMPLATE

The artwork template (shown in Figure 10-1) developed by the Motion Picture and Education Markets Division of Eastman Kodak Company, provides a convenient framework within which to design graphs and tables. Three sets of outlines can be used, one for single-frame filmstrips with 16 millimeter motion pictures, the other for 35 millimeter lantern slides, and the final one for television images.

The template is based on the standard proportions of screen size to auditorium or theater dimensions most widely used in the United States and most other countries. The sample letters "A" and "b" given refer to the viewing distance in relation to screen width. For example, if the screen width is 5 feet, then the viewing distance at which the sample letter can normally be seen by the audience is "6W." This means "6 times the screen width" or $6 \times 5 = 30$ feet in our example. For other ratios, such as 4W and 12W, different letter sizes are applicable.

LETTER SIZES

Viewing distance *(factor times width in feet)*	*Minimum Size of letter* *or symbol (inches)*
4W	3/32
6W	1/8
8W	5/32
10W	3/16
12W	1/4

Viewing distance *(factor times width in feet)*	*Divisor to find* *size of lettering*
4W	75
6W	50
8W	35
10W	30
12W	25

Sizes of letters and symbols are given in terms of height, excluding ascenders or descenders (such as the rising tail of "b" or the dragging tail of "p"). Naturally, the smallest letter used governs the overall legibility of the material. The dimensions are, moreover, minimal. Larger sizes are preferable, if space permits. Minimum line weights are governed by standard pen sizes. For this purpose, up to 6W of distance permits use of a No. 00 pen, while from 8 to 12W a No. 1 pen is applicable.

SPECIAL FORMATS

On certain occasions, special formats may be required. This occurs whenever screen and theater dimensions do not conform to standard proportions. For such instances, the template shown in Figure 10-2 is used. The letter sizes are now given by division of the viewing distance factor, as follows:

If the height of the art work information area is 11 inches and the viewing distance is to be 4W, divide 11 by 75. The result is .146, which is slightly larger than 9/64 inch. It is the minimum letter size to be used.

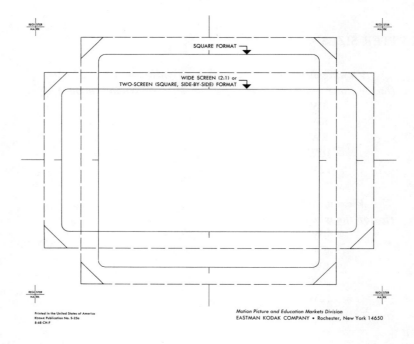

Printed in the United States of America
Kodak Publication No. S-25a
8-68-CH-F

Motion Picture and Education Markets Division
EASTMAN KODAK COMPANY • Rochester, New York 14650

Figure 10-2: Art work template for nonstandard (square and wide) screens. Template courtesy of Motion Picture and Education Markets Division of Eastman Kodak Company, Rochester, N.Y.

The information provided above is based on a (noncopyrighted) pamphlet, "Kodak Artwork Size Standards for Projected Visuals," prepared by the Motion Picture and Education Markets Division, Eastman Kodak Company, Rochester, New York 14650 (Kodak Pamphlet No. S-12, 10-66 Major Revision, 266-L-RPP-AX).

USE OF A TYPEWRITER

By far the most available, handy, and convenient way of lettering is by means of a typewriter. For this purpose, whether it be in adding type to diagrams or just in the preparation of tables for projection, the template in Figure 10-3 will be of considerable value.

In effect, this template reduces the overall information area and thereby increases the proportionate size of typed material. Straight typing of the desired material will now produce a good, readable size for projection. Pica-sized type is preferable, while elite type is barely acceptable. When

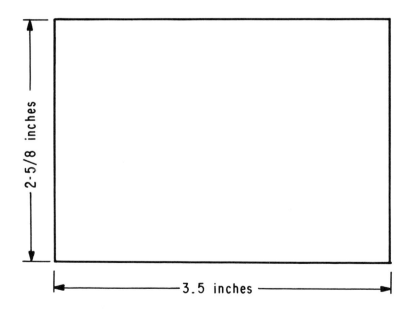

Figure 10-3: Lantern slide template. Each of the boxes shown is in proper proportion (the so-called aspect ratio) for the standard 35 millimeter lantern slide. The dimensions of 3½ times 2⅝ inches are such that direct typing (pica size) will result in a properly readable text using screen projection for an audience of 50 to 150 people.

elite must be used, one should keep in mind that inscriptions with all capital letters are more readable at a distance than upper and lower case text.

TELEVISION SCREENS

From a review of the Kodak art work template, it will be noted that the permissible dimensions for television titles are narrower than those for 35 millimeter or motion picture projection. The decrease in dimensions, and thus relative increase in the size of the writing, allows for lack of distinctness in transmission and some degree of image distortion and blurring that may occur.

TEN-INCH TRANSPARENCIES

Ten-inch-wide transparencies, for use in the ordinary overhead projector, are prepared in the same manner as 35 millimeter slides. From this art work, the development of either a lantern slide or transparency is an optional matter. Both will be equally visible and legible at a given distance. The reason for this is that it is the size of the detail on the screen, and hence the size of lettering and lines to the overall art frame dimensions, that is important and governing.

INTERCHANGEABILITY OF USE

It will be noted that art work for projection or tables for projection can be readily used in reports destined for close-at-hand reading. But the reverse need not be true. Reports permit much smaller type with perfect legibility.

When material is to be converted from report to projection use, the following problems will be involved:

1. Check letter and other detail sizes. If they fit the requirements for good projection, all is well. If not, revision will be needed.
2. One common way to convert materials to projection use is to omit detail. This detail is then supplied by the speaker during his talk. It is often a good idea to distribute copies of the talk to the attending group.

3. Another approach is to convert a single diagram or table into a number of separate diagrams or tables. The material so dissected is then given sequentially. There is some loss of overview in this technique. A single chart may bring together many aspects or concepts applying to a certain presentation. When this material is "cut up," comprehensiveness may be lost. Again, of course, supplementary printed copies of the paper may be supplied to the audience.

While in theory it seems reasonable that one diagram or one table be used for many purposes, the practical aspects of medium, manner of presentation, and type of audience may call for several versions of the same diagram or table. The extra work that is called for should not be done grudgingly, if the aim of purposeful, informative, and persuasive presentation is to be attained.

11. Preparation of Source Credits

Material on charts and tables need not represent the author's single-handed effort.* Many times he will have resorted to, reused, or adapted prior data, and similarly new ideas may have been structured on a foundation of work and concepts developed by others. Source credits must then be given, whether copyrighted or other proprietary material is used with permission, or subject matter and data in the public domain are taken. In the latter instance, there is usually an ethical obligation.

The degree of formality in giving credit for sources varies with the type of writing. A research-scientific report usually calls for detailed citing of sources so as to facilitate review of the sources by others. The procedures and style for source credits provided here represent relatively detailed and formal description, with emphasis on consistency of usage. Where lesser detail or formality seems desirable, appropriate modifications should be applied.

ABBREVIATIONS

Common abbreviations used are:

Full word	Singular	Plural
page	p.	pp.
volume	vol.	vols.
number	no.	nos.
chapter	chap.	chaps.
editor or edition	ed.	eds.

The abbreviations are usually followed by a period.

* This chapter was coauthored by Mrs. D. L. Lewis, whose research in general usage was underwritten by the Center for Business and Economic Research of Kent State University.

BOOKS

The general order of information given for a book is as follows:

1. Name of the author in the form appearing on title page with the surname first. That is, if the name on the title page is David Leslie Jones, do not abbreviate it to David L. Jones or D. L. Jones. Give the last name first. End the name with a period.

2. The title of the book and the subtitle if one is given. This is underlined, which indicates italics to a printer. First letters of main words are capitalized. This is also followed by a period. (A different form followed by some is to capitalize only the first letter of the first word in the title.)

3. The number of the edition if it is other than the first. The period ending the abbreviation for edition serves as end punctuation.

4. The number of volumes is given if there is more than one volume. This is followed by a period.

5. The place of publication. It is common to use New York for New York City, New York, or Chicago for Chicago, Illinois. Lesser known locations are listed by both town and state, or in the case of foreign cities not widely known, the city and country are given. This is followed by a colon.

6. The name of the publisher. Again, some shortening of the name is common. For example, a well-known firm such as McGraw-Hill Book Company is usually written as McGraw-Hill, but Gramercy Publishing Company is not shortened. Words such as "company" and "incorporated" are abbreviated. This information is followed by a comma. If no publisher is indicated, write "n.p."

7. The date of publication. This is the date on the copyright page. If there is no copyright, use the date on the title page. What is needed is the year in which the book was published or the year of the most recent revision. A book first published in 1960 but which has been printed a number of times since then, but not revised, still is listed with 1960 as the correct date. An older book that has been reissued by different publishers, perhaps in both hardback and paperback, would be cited by the date and publisher of the edition the researcher used, followed by the date the book was first published. "First published in . . .", followed by the year, would be the form to use. If no date can be found, write "n.d."

8. The number of pages is the last part of the entry and is followed by a period. The form used here is to record the number on the last numbered page of the book, followed by the abbreviation for page. Some publishers and style manuals record the facing pages which are numbered in Roman, "pp. xiv, 321," for example. Still another variation is not to give the number of pages at all.

PERIODICALS

The general order of information given for an article in a journal is as follows:

1. Name of the author, surname first. Again, record the name exactly as it is written in the source. The name is followed by a period.
2. Title of the article in quotation marks, with the closing period within the quote marks. The first letter of main words in the title should be capitalized. Some style manuals do not capitalize any letter beyond the first letter of the first word. Another matter of style concerns the end punctuation, which is sometimes a comma.
3. The name of the periodical in which the article appeared. This is underlined and normally followed by a period. (Again a different style is to always use a comma after the journal name.)
4. The volume of the periodical. The volume is given in arabic even if roman is used by the journal. The abbreviation for volume precedes the actual number. There is no end punctuation, since parentheses follow next, enclosing the date. If the exact date is given (month and year, or month, day, and year) identifying a particular issue, mention of the issue number may be omitted. However, if only the year is given, but several issues appear throughout the year (and are not consecutively paged), the issue number becomes very important. If there is no volume number, the issue number is also very important. (Variations of style here include giving the issue number of the journal in addition to the volume, such as vol. 52, no. 3. Another style sometimes followed is to use roman numerals for the volume. Still another variation is to write only the actual volume number, eliminating the word, volume.)
5. The date of issue is given in parentheses, followed by a comma.
6. The last item given provides the inclusive number of the pages on

which the article appears. The abbreviation for pages precedes these numbers, and a period ends the entry. (Logic dictates the placement of the abbreviation for pages as one says the word before the numbers. For example, "It is on pages 12 to 22." Whereas the usual means of expression for a total number of pages in a book is, for example, "It is 350 pages in length." Hence, the variation in placement of the abbreviation for pages in a bibliographical entry.) Another form sometimes used gives the actual numbers only, eliminating the "pp." Nonsequential paging is handled as for newspapers (see below).

NEWSPAPERS

The form of entry for a newspaper is the same as the form for an article in a journal, except that it is shorter, because the volume, number, and (except for the *New York Times* and London *Times*) page numbers need not be given. The specific date is enough information for easy location of any daily, weekly, or monthly publication. If no author's or reporter's name is given, the article is alphabetized by the first main title word along with any other unsigned pieces after all signed materials have been listed.

THE BIBLIOGRAPHY

The order in which bibliographic entries are given will depend on the variety of sources used. For lengthy bibliographies, it is suggested that books be listed first, as one section, followed by articles. The section of articles, if lengthy, can be divided into journal articles and newspaper and other sources. Pamphlets and reports can be alphabetized with the books, but if such documentation is numerous it would be better to have a separate section for them. Interviews, letters, and patents might each be in a separate section. Any miscellaneous sources can be combined in a section headed as "Other Sources" or "Miscellaneous Sources."

EXAMPLES OF BOOKS

Books by one author

DOMMERMUTH, WILLIAM P. *The Road to the Top: A Study of the Careers of Corporation Presidents.* Austin, Texas: University of Texas, Bureau of Business Research, 1965. 128 pp.

HOROWITZ, IRVING LOUIS. *Three Worlds of Development: The Theory and Practice of International Stratification*. New York: Oxford University Press, 1966. 475 pp.

KUZNETS, SIMON. *Modern Economic Growth: Rate, Structure, and Spread*. New Haven, Conn: Yale University Press, 1966. 529 pp.

MAMBERT, W. A. *Presenting Technical Ideas: A Guide to Audience Communication*. New York, N. Y. 10016: John Wiley & Sons, Inc., 1968. 216 pp. Emphasizes methods of preparation and delivery of technical papers.

More than one book by the same author. These are listed alphabetically by the first main title word. After the first entry, a dash indicates the same author.

GALBRAITH, JOHN KENNETH. *Economic Development*. Cambridge, Mass.: Harvard University Press, 1964. 109 pp.

————. *The Affluent Society*. Boston: Houghton Mifflin Company, 1958. 368 pp.

Edited books (that is, a book compiled from other published material).

GROSS, BERTRAM M., ed. *Action Under Planning: the Guidance of Economic Development*. New York: McGraw-Hill Book Company, 1967. 314 pp.

HAUSER, PHILIP M., AND LEO F. SCHNORE, eds. *The Study of Urbanization*. New York: John Wiley & Sons, Inc., 1965. 554 pp.

Books by more than one author

Note that the second author's name is not inverted, as it is not being alphabetized. Also note the comment in parentheses to clarify the date of publication of the material in the following entry.

BETTLEHEIM, BRUNO, AND MORRIS JANOWITZ. *Social Change and Prejudice: Including Dynamics of Prejudice*. New York: The Free Press, 1964. 337 pp. ("Dynamics of Prejudice" was originally published in 1950 and comprises two-thirds of the book.)

WILSON, G. W., B. R. BERGMANN, L. V. HIRSCH, AND M. S. KLEIN. *The Impact of Highway Investment on Development*. Washington, D. C.: Brookings Institution, 1966. 226 pp.

Note: Always list the names of authors in the order in which they appear on the book, which may be alphabetized or may have as the first author's name the person most responsible for the book.

Books, no author listed. Such books are included in alphabetical order by title after all authored books have been listed.

Anticipating the Nation's Needs for Economic Knowledge. 46th annual report, June 1966. New York: National Bureau of Economic Research, 1966. 92 pp.

Book, translated

SIMONNARD, M. *Linear Programming.* Translated by W. S. Jewell. Englewood Cliffs, N. J.: Prentice-Hall, 1966. 430 pp.

Books in volumes

LEOPOLD, R. W., A. S. LINK, AND S. COHEN, eds. *Problems in American History.* 3rd ed. Vol. I, *Through Reconstruction;* Vol. II, *Since Reconstruction.* Englewood Cliffs, N. J.: Prentice-Hall, Inc., 1966. 873 pp. Note: both volumes are listed only if both are used.

MEADE, J. E. *Principles of Political Economy.* Vol I: *The Stationary Economy.* Chicago: Aldine Publishing Company, 1965. 238 pp.

Corporation author

International Business Machines Corporation, *IBM 709 Data Processing System.* White Plains, N. Y., 1960. 174 pp.

Note that here no publisher is given, as none was listed. The publisher is assumed to be IBM itself, which does not need to be repeated after the home office location.

General Motors Corporation Research Laboratories. *Symposium on Adhesion and Cohesion.* Warren, Mich., 1961. Edited by Philip Weiss. Amsterdam, N. Y.: Elsevier Publishing Co., 1962. 272 pp.

Note that in this entry, although there is an editor listed, he is secondary to the sponsor, the corporation, involved.

Book (or pamphlet) with an association as author and publisher

National Industrial Conference Board. *Population and Economic Growth: A Chart Guide.* New York, 1966. 22 pp.

(Note: A variation in style in presenting such an entry would be to rename the author after the place of publication. It could be abbreviated at that point to NICB, as could have been done with the IBM entry just cited.)

Yearbooks

It is not necessary to give the number of pages for such books.

U.S. Bureau of the Census. *Statistical Abstract of the United States: 1967.* 88th ed. Washington: U.S. Government Printing Office, 1967.
United Nations. *Statistical Yearbook, 1966.* New York: U.N. Statistical Office, 1967.

> Note: For less well-known United Nations documents, it is helpful to give the sales number, to facilitate locating or ordering. This number is found within the first pages of a United Nations document. It should be given in parentheses immediately after the year of publication.

U.S. Department of Labor, Bureau of Labor Statistics. *Handbook of Labor Statistics, 1967.* Washington, D. C.: U.S. Government Printing Office, 1967.

PATENTS

The order of a patent citation is as follows: the patentee, followed by the name of the assignee (in parentheses) if there is one; the title of the patent; the patent number, serial number, and date of filing or issuance.

BINDERMAN, WALTER. Extendable Wick Candle. U.S. Patent No. 3,360,966, serial no. 594,801. Filed Nov. 16, 1966.
FLOWER, ROBERT A. AND GUS STAVIS (to General Precision, Inc.). *Optical Radar System.* U.S. Patent No. 3,360,987, serial no. 380,156. Filed July 3, 1964.

> Note: If the patent title does not communicate its function, the title should be supplemented. The *Official Gazette* of the U.S. Patent Office contains explanatory paragraphs that can provide the needed additional information.

Theses

The form for citing a thesis in a bibliography is as follows:
CARNAHAN, GEORGE RICHARD. "Organizational Stability in Manufacturing Concerns in the United States." Ph.D. dissertation, Ohio State University, 1967. 156 pp.
SUGG, HOWARD A. I. "Soviet Disarmament Theory Since 1959: An Analytical Study." Ph.D. dissertation, The American University, 1967. 573 pp.

PERIODICALS

BELASCO, JAMES A. "The Salesman's Role Revisited." *Journal of Marketing* 30 (April 1966): 6-11.

BURCK, GILBERT. "A New Business for Business: Reclaiming Human Resources." *Fortune* 77 (January 1968): 159-61, 198-200, 202.

HARSANYI, JOHN C. "Games with Incomplete Information Played by 'Bayesian' Players, I-III. Part I, the Basic Model." *Management Science* 14 (November 1967): 159-82.

MORSS, ELLIOT R., J. ERIC FREDLAND, AND SAUL H. HYMANS. "Fluctuations in State Expenditures: An Econometric Analysis." *Southern Economic Journal* 33 (April 1967): 496-517.

SCHULTZ, THEODORE W. "Urban Developments and Policy Implications for Agriculture." In *Economic Development and Cultural Change* 15 (October 1966): 1-9.

WEINSTEIN, EUGENE A., MARY GLENN WILEY, AND WILLIAM DE VAUGHN. "Role and Interpersonal Style as Components of Social Interaction." *Social Forces* 45 (December 1966): 210-216.

The next two examples are unsigned articles which are alphabetized after signed material has been listed.

"How FTC Keeps up on Mergers." (Computer points out trends and will show directions that companies are going.) *Business Week* no. 2021 (May 25, 1968): 132-133.

Note that further explanatory material was given in parentheses to tell the reader more about the article. No volume number appears, since *Business Week* uses only an issue number.

"Student Power." *New Republic* 158 (May 25, 1968): 11-12.

Articles in a newspaper

ALLAN, JOHN H. "After Skirting Brink, Tax Game Nears End." *New York Times,* June 2, 1968, Sec. 3, pp. 1, 7.
Note that in citing newspapers no parentheses are used around the date because no volume number need be given.

GOVERNMENT DOCUMENTS

Government report. The form used here involves first listing the name of the country, state, city, or county involved, followed by the name of the agency, if one was given, and then proceeding to the title and other data.

Cuyahoga Falls, Ohio, Cuyahoga Falls City Planning Commission. *Annual Report on Planning, 1961.* Cuyahoga Falls, Ohio, 1962. 27 pp.

U.S. Senate, Committee on Finance, *Unemployment Insurance Amendments of 1966: Report together with minority views to accompany H. R. 15119.* 89th Cong., 2nd sess. Finance Committee report no. 1425. Washington, D. C.: U.S. Government Printing Office, 1966. 92 pp.

Government serial. The United States government publishes many journals and magazines. In order to assist the reader in identifying and finding unfamiliar material, it is helpful to indicate the source of the publication. Libraries frequently catalog and store government publications separate from other documents. A journal such as the following example is more quickly found if the government source is noted.

BITTERMANN, HENRY J. "Treasury Reporting on Foreign Grants, Loans, and Credits. Statistical Reporter. No. 68-10, April 1968. (Washington, D. C.: Bureau of the Budget), pp. 161-66.

PERSONAL INTERVIEW REFERENCES

WARREN, THOMAS S. Sales Manager, Strong Steel Co., Lausdale, Ohio. Personal interview on May 20, 1968 regarding promotion criteria.
Note: If the researcher feels that a summary of the interview is pertinent and it has not been covered in the text of the paper, then it should be added in an appendix. This is also true for references to letters. Each interview or letter comprises a separate appendix section and is mentioned in the bibliographical entry.

LETTER REFERENCES

BOWMAN, WAYNE M. City Manager, Lincoln, Ohio. Letter regarding zoning practices, dated April 12, 1968. See Bibliography.
Those who wish a more extensive treatment of source credits should see the 12th edition of *A Manual of Style,* published by the University of Chicago Press; the stylebook of the U.S. Government Printing Office; and *The New York Times Stylebook for Writers and Editors,* published by McGraw-Hill Book Company.

12. Forms Design

Charts and graphs are derived from data. This data, in turn, comes from a variety of sources. In order to acquire systematic, comparable, and complete items from which appropriate summaries can be developed, a standardized method for acquiring data inputs—the questionnaire or recording form—is often the most practical and effective approach.

DESIGN PRINCIPLES

Most of the prescriptions for the development of charts, graphs, and tables apply also to forms. For example, primary emphasis should be toward the development of relevant and important information. The types of data to be entered should be clearly identified. Careful design, leaving adequate space to supply requisite information entries, combined with instructions that indicate clearly the types of data desired—all will help in attaining the cooperation of the person completing the form.

In recent years there has been a strong trend toward box design in forms development. As a result, the information to be supplied for a particular section of the form is "bordered off." An example is provided in Figure 12-1, where each reporting item is clearly defined in terms of a rectangular box. Experience has shown that the person filling out the form is encouraged to be concise and to stay within the prescribed limits. In subsequent analysis and evaluation of accumulated forms, the box design facilitates the summarizing of information by various categories and types.

An alternative to the boxed design is the open form, in which the limits for the questionnaire response are ill-defined. While the open form continues to be widely used, it represents a less structured approach.

161

OFF-STANDARD REPORT								
Mill:	Dept.:				Date of Test:			
Type of Test:			Std.:		High:		Low:	
Mach. No.:								
Item								
Test Result								
Check Test								
Date:								
Date Submitted:	by Lab. Supervisor				Work Completed Date: Overseer			
Notes:								

Figure 12-1: Form showing boxed-entry design.

In many instances, a combination of boxed and open form may be desirable. The procedure used then is:

1. The first part of the form calls for the providing of simple, direct, explicit information. The boxed design is most appropriate for this purpose.
2. The second part provides for more general responses, open-ended replies, statements of opinions, reports of particular conditions, information not amenable to short answers, or other such materials.

FUNCTIONS OF FORMS

The function of a form represents the purpose that is to be accomplished. Often, several functions are involved. Among them, we may note such functions as acknowledgments, authorizations, applications, cancellations, follow-ups, notifications, recording and reporting, requests, routings, and schedulings.

In the following are illustrations of forms serving a variety of functions.

- An application for employment is shown in Figure 12-2. The boxed form is used, but it is clear that additional information may be provided on separate pages in an open-ended manner.
- A segment of a cost-estimating form is provided by Figure 12-3. At least in part, this form is a tabulation. Uniform cost calculations are assured by this form.
- Forms that fit standard dimensions of data processing equipment are becoming quite popular. A typical example, representing a stores requisition, is given in Figure 12-4.

Usage of forms is likely to increase rapidly in the future, precisely because we are relying more and more on automated data processing most effectively and efficiently when they are coupled with unified, systematic, and well-coordinated inputs. As a result, the computer becomes a "disciplinarian" calling for uniform, consistent types of information inputs—and well-designed forms are a means of conforming to these requirements.

CONCLUDING OBSERVATION

Forms are a communication tool whereby information is gathered, compiled, summarized, analyzed, and evaluated. This function and usage can in many ways be integrated with and overlap those activities that lead to charts and graphs and tabular presentations. It is for these reasons that the present brief chapter on forms design has been included.

resume

0 — Me. Vt. N.H.   1 — N.Y. Pa. Del.
Mass. R.I.   2 — Md. W. Va. Va.
Conn. N.J.   N.C. S.C.
Virgin Islands
Puerto Rico   9 — Alaska Hawaii
(not shown)   (not shown)

1. PERSONAL DATA (Please Print or Type)

Membership Number	Sex/Marital Status	Year of Birth	Dependents	Citizenship
	☐ —Male Single ☐ —Male Married ☐ —Female Single ☐ —Female Married			☐ —U.S. Citizen ☐ —Non Citizen ☐ —Permanent Visa

Military Service Status	Security Clearance
☐ —Active Service Completed ☐ —Not Eligible ☐ —Eligible or Active Service Not Completed	☐ —Confidential ☐ —Top Secret ☐ —Secret ☐ —Atomic Energy Commission

2. SALARY PROGRESS

GEOGRAPHIC PREFERENCE

Total Monthly Earnings	Amount of	Date of Last Incr.		Amount of	Date of Prev. Incr.		Monthly	Any	No	Mark the first two or select up to 4 areas
	Last Increase	Mo.	Year	Previous Incr.	Mo.	Year	Base Salary	area move		0 1 2 3 4 5 6 7 8 9

3. EDUCATIONAL BACKGROUND

Highest Level	Year 19___ Graduated	Name of School
Major field of Highest Level		Additional Field of Study

4. PROFESSIONAL EXPERIENCE PROFILE

Current Industry Experience	M31	General Occupational Field	Years of Experience

SKILLS: Consult the reverse side of this form for skill codes applying to your specific experience. Enter up to 14 codes, giving approximate number of years for each skill. If you do not find an appropriate code listed, enter your description and leave code area blank.

Code	Yrs. Exp.	Specific Professional Experience	Code	Yrs. Exp.	Specific Professional Experience

5. EMPLOYMENT HISTORY

Name of Company/Parent Company	Position Title	From		To		Monthly Salary
		Mo.	Year	Mo.	Year	
Description of Duties						

Previous Employer	Position Title	From		To		Monthly Salary
		Mo.	Year	Mo.	Year	
Description of Duties						

Second Previous Employer	Position Title	From		To		Monthly Salary
		Mo.	Year	Mo.	Year	
Description of Duties						

Figure 12-2: Forms design, further illustration.

PART III - COST AND PRICE ANALYSIS - RESEARCH AND DEVELOPMENT CONTRACTS *(The information requested below must be complete when submitted with proposals for the procurement of research and development services. If your cost accounting system does not permit analysis of costs as required, contact the Purchasing Office for further instructions.)*

DETAIL DESCRIPTION	ESTIMATED HOURS	RATE/HOUR	TOTAL ESTIMATED COST (Dollars)
1. DIRECT LABOR *(Specify) (See Part VI - Biographical Sketches)*			
TOTAL DIRECT LABOR			

2. OVERHEAD COST ON DIRECT LABOR ABOVE	OVERHEAD RATE X BASE = OVERHEAD ($)		
TOTAL OVERHEAD			

3. OTHER DIRECT COSTS: *(Specify in Exhibit B on Page 3 for additional space)*		EST. COST ($)	
TOTAL OTHER DIRECT COSTS			

4. SUBCONTRACTS *(Specify in Exhibit A on Page 3)*			
TOTAL SUBCONTRACTS			

5.	TOTAL DIRECT COST AND OVERHEAD	
6. GENERAL AND ADMINISTRATIVE EXPENSE *(Rate % of Item Nos.*		
7.	ESTIMATED COST	
8. FIXED FEE/PROFIT *(State basis for amount in proposal)*		
9. SPECIAL EQUIPMENT *(Specify in Exhibit B on Page 3)*		
10. TRAVEL *(If direct charge)*		
A. TRANSPORTATION		
B. PER DIEM OR SUBSISTENCE		
TOTAL TRAVEL		
11. CONSULTANTS *(Identify - purpose - rate)*		
TOTAL CONSULTANTS		
12.	TOTAL DIRECT REIMBURSEMENTS	
13.	TOTAL ESTIMATED COST AND FIXED FEE/PROFIT	

Figure 12-3: Forms design continued.

165

3 2 T 1 7 4 U 2 1 7 4 T 2 1 7 4 U 2 1 7 4 H 2 1 7 4 T 2 1 7 4 U 2 1 7 4 T 2 1 7 4 U 2 1

DIVISION		SECTION	STORE CREDIT (NAME)	ALLOTMENT CODE NUMBER	EXPENDITURE CODE

DATE

157340

| ISSUED TO | | | APPROVED BY | |
| ISSUED BY | DELIVERED TO | RECEIVED BY | POSTED TO RECORD BY |

DISTRIBUTION CODE	APPROP.	DIVISION	SECTION	ALLOTMENT	EXPENDITURE CODE

QTY. ON HAND	QTY. REQUISITIONED	DESCRIPTION	STOCK NO.	QTY. DELIVERED	UNIT PRICE	AMOUNT

APPROPRIATION

←——— T O T A L S ———→

STORE REQUISITION — STATE OF LOUISIANA

Figure 12-4: Form that fits standard dimensions of a data processing system.

13. Epilogue: Data Analysis and Communication—Today and Tomorrow

"The higher one looks in administrative levels of business, the more one finds that decisions are based on data that are analyzed statistically and presented in tabular or graphic form."

This observation by Charles A. Bicking* underscores the crucial place of information as a decision-aiding and communications factor at the higher echelons, not only of business, but for all types of organizations. One might add that these communicative aids also serve in other levels of decision-making, and in fact, throughout the entire hierarchy of management goals, planning, action, and control. Furthermore, flow charts and related types of diagrams are needed to portray and review relationships and operational interfaces that are involved in both large-scale and modular components of systems and procedures applications.

* Dr. Bicking is manager of the Mathematics Branch, Research and Development Division of the Carborundum Company. The quotation comes from a talk by him and was also used by him in his subsequent contribution to N. L. Enrick, *Cases in Management Statistics,* New York: Holt, Rinehart & Winston Inc., 1962.

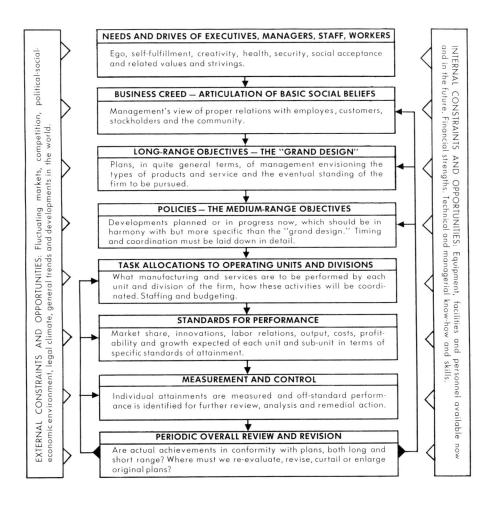

EXTERNAL CONSTRAINTS AND OPPORTUNITIES: Fluctuating markets, competition, political-social-economic environment, legal climate, general trends and developments in the world.

NEEDS AND DRIVES OF EXECUTIVES, MANAGERS, STAFF, WORKERS

Ego, self-fulfillment, creativity, health, security, social acceptance and related values and strivings.

BUSINESS CREED — ARTICULATION OF BASIC SOCIAL BELIEFS

Management's view of proper relations with employes, customers, stockholders and the community.

LONG-RANGE OBJECTIVES — THE "GRAND DESIGN"

Plans, in quite general terms, of management envisioning the types of products and service and the eventual standing of the firm to be pursued.

POLICIES — THE MEDIUM-RANGE OBJECTIVES

Developments planned or in progress now, which should be in harmony with but more specific than the "grand design." Timing and coordination must be laid down in detail.

TASK ALLOCATIONS TO OPERATING UNITS AND DIVISIONS

What manufacturing and services are to be performed by each unit and division of the firm, how these activities will be coordinated. Staffing and budgeting.

STANDARDS FOR PERFORMANCE

Market share, innovations, labor relations, output, costs, profitability and growth expected of each unit and sub-unit in terms of specific standards of attainment.

MEASUREMENT AND CONTROL

Individual attainments are measured and off-standard performance is identified for further review, analysis and remedial action.

PERIODIC OVERALL REVIEW AND REVISION

Are actual achievements in conformity with plans, both long and short range? Where must we re-evaluate, revise, curtail or enlarge original plans?

INTERNAL CONSTRAINTS AND OPPORTUNITIES: Equipment, facilities and personnel available now and in the future. Financial strengths, technical and managerial know-how and skills.

Figure 13-1: Hierarchy of management goals. Within the context of management planning and control, as depicted in the eight successive plates above, goal-setting takes various forms. Needs and drives of individuals, business creeds, long-range and medium-term objectives, task allocations, and standards all represent descending hierarchies of goals governing the activities of an organization. Measurement, control, and review lead to eventual revision of goals, allowing for internal and external constraints and opportunities. From N. L. Enrick, *Decision-Oriented Statistics,* Princeton, N.J.: AUERBACH Publishers Inc., 1970.

The nature of the hierarchy is further elaborated in Figure 13-1, which also notes the confines of external and internal constraints and opportunities that bear on the knowledge and information gained. For the formation of the business creed, the long-range objectives, the policies and task allocations, there must be awareness not only of the pertinent facts, but also of significant relationships; potential future conflicts, inconsistencies, and divergencies; and flows of materials, processes, procedures, and information, so that major and minor decisions will be workable and in harmony with the total goals of the complex. Both qualitative and quantitative facts enter into the process. Information must be visualized in compact, comprehensible perspective, thereby facilitating the arrival at proper conclusions, the study of pertinent alternatives, and the coming up with useful structures that lead to effective problem solutions.

Another facet of the management goals hierarchy involves the setting of performance standards, the measurement of performance, establishment of control limits, and periodic overall review and revision of policies and objectives, standards, and controls. Here again, data and relationships analyzed and depicted in proper perspective, highlighting the essential findings and correlations, are the best basis on which to proceed. The context in which this work is carried on is further elaborated in a number of illustrative diagrams that have been presented in earlier chapters. The reader may be particularly interested in reviewing the following:

- Weighing of alternatives and limitations on alternatives, as in Figure 5-3.
- Procedures establishment, as in Figure 5-5.
- Model-building for systems design and analysis, as in Figure 5-6.
- Policy formulation, as in Figure 5-7.
- Curriculum-building, as in Figure 5-15.
- Organization chart establishment, as in Figure 5-16.
- Consideration of alternative organization patterns, as in Figure 5-17.
- Multiple decision problems and problem-solving, as in Figure 6-3, supplemented by the generalization in Figure 6-4.

While these diagrams were presented from the primary viewpoint of showing charting and graphing methodology, they also, as an incidental by-product, reflect the essential elements that go into the systematic, quantitatively oriented management process. Nor is this process confined to business and industry. Government, research groups, professional associations, and other groups will usually, at least in some form, use similar or quite parallel managerial and goal-setting applications. Good practice in planning a marketing strategy involves principles and procedures that are also good practice in planning a research project in the physical, biological, or social sciences. In each case there are goals, there are limitations on alternative strategies of attack on the problem, and there is need to review data, relationships, information, and observations so as to enhance the likelihood of successful decisions and outcomes.

It must be recognized that providing the right information is itself a management problem requiring proper analysis and resolution. All too often we find that there is neglect of this phase, with consequent detrimental effects. For example, the goal of a firm may be threefold: (1) to produce at minimum cost, (2) to maintain requisite quality and reliability of product, and (3) to meet promised delivery schedules. Yet good information was provided only on the first objective. Not only were the other two important objectives shortchanged as a mere result of data inadequacy, but conflicts—such as between cost and quality, or quality and delivery—could not be resolved optimally.

Looking toward tomorrow, we perceive an accelerating progress of science and technological innovation, a quickened pace of changes throughout the fabric of society, and a proliferation of the flow of ideas and ideologies that are brought to bear on us. Merely from the standpoint of keeping up with relevant information alone, there are major problems. But there is the additional need of interfacing areas of work. Managers, scientists, technologists, mathematical analysts, and computer programmers must relate well to each other and, in the interface of the problem areas, work together in coming up with proper responses to needs and threats, potential pitfalls, and promising opportunities.

Future computer and data processing systems will be vastly superior to present models. First of all, we expect better software—the programs that are available to solve complex, interlocking data structures and discover meaningful relations and relative costs and cost effectiveness. Next, hardware will be improved as regards costs per output of information. This will be accomplished by intense further microminiaturization and modular buildup of computer circuits, produced by highly automated equipment. Designers talk in terms of "one-hundredth of present cost" rather than "one-tenth of present cost," in discussing circuits and components that theoretically are feasible and on which practical work is now being done. By the use of data sharing, a firm of any size and in any type of business will be able to utilize the computing abilities of automated data processing equipment. Moreover, the benefits of data storage, data retrieval, and rapid calculations will become available to anyone—in the sciences, the arts, and other endeavors.

While efficient and reliable automatic data-plotting equipment is available today, the output may be restricted by limitation on types of print, shading of lines, and working-in of detail. As a result, the overview value of graphs and charts suffers. There are similar objections to tabular material. In the years to come, it may be anticipated that data-plotting and data-tabulating equipment will become available, which will go a long way toward overcoming present limitations. Moreover, with proper software, it should be possible to have automatic transfer from data columns to charts of a chosen design. A well-prepared chart, which today may have consumed a considerable amount of time in planning, reworking, and final drafting, may be produced by the computer and auto-plotter in a matter of minutes. Again, let it be emphasized that we already have such equipment, but from many viewpoints, it is usually inferior to personally designed charts. There is, of course, no denying the incomparably greater speed and labor savings of automated graphing. For this reason, automated graphing and charting will find a manifold increase in applications in the years to come.

Still another aspect of data processing should be mentioned. It concerns the need for data parsimony in the final output. So thunderous has been the outpouring of data floods in many information systems, that engulfed humans have been calling for a halt. The choice between wading through miles of computer print-out for needed information, or ignoring available data, is not a happy one. Rescue, it appears, may come from the efforts of people developing methodologies for data parsimony. Concern is with the collection, enumeration, classification, and consolidation of data in a manner that will relay crucial facts, ready overviews, and quick insights parsimoniously though adequately. Control chart methodology, such as was illustrated in Figure 3-10, holds a key approach in this regard. As a production-management tool, for example, it involves the following thought processes: because of some variability in materials, processes, measuring techniques, and ambient conditions, proper control of quality distinguishes between "chance" and thus allowable variations in dimensional strength, or other characteristics of output and significant or findable causes of variation. The pertinent distinguishing criterion is the control limit. When actual test results, plotted on the chart, fail to stay within allowable regions, this fact signals the probable presence of trouble in the processing equipment, the material being fed, or the method of operation. Corrective action is thus called for. We recognize in the control chart an important principle of good management.

Do not concern yourself with minor variations in performance. Concentrate, instead, on real and significant deviations from goal levels. Permit the multitude of insignificant detail, pertaining to a given operation, to remain in the background. It will stay there—except when a control limit has been exceeded.

All routines of management information data, it seems, can be placed on this principle. Relevant procedures, concepts, and applications of control limit criteria can thus be made an integral part of quantitatively oriented management. Moreover, usage should not be spotty, isolated or in bits-and-pieces terms, but as a total systems contribution.

But application of the exception principle does not represent the only area in which data parsimony can be achieved. For example, in Figure 4-2 we showed how an entire pattern of variability can be expressed in terms of a single value, the coefficient of variation. Now let us say a firm desires to examine the data pattern of a group of manufacturing plants, sales offices, or other items. One way would be to look at the frequency distribution graphs for each of the plants or offices. But if there are many such graphs, the people concerned will soon feel flooded and overwhelmed by the information, particularly if numerous categories of operation and variations are thus to be studied. Instead, we may compute variation coefficient data for each frequency distribution pattern. Next, a system may be devised whereby a single graph contains a total information result for comparative studies and interplant or interoffice comparative evaluation of average performance and variation in performance.

We have given examples of known techniques for data parsimony. These permit abstraction of the essentials from a whole complex of information detail, without omitting any vital aspects, but furthering the decision processes of management. In the future there will be creation of an entire array of new techniques and systems that place further demands on human perceptive and cognitive powers. Again, computer software will have to be developed which will automatically convert and transform data to desired form and which will abstract essentials according to chosen patterns. Much of this is being done today.

A strong point can be made for the argument that data parsimony is hindered by the fact that managers do not seem to be able to spare the time and effort needed to understand compressive information parameters. For example, the variation coefficient previously discussed is a compressive term. Within one value, a whole range of variability is captured. For example, other data having a higher coefficient are then recognized as exhibiting a wider range of variation. Further illustrations can be given. Thus the correlation coefficient in Figure 4-6 expresses in one figure the degree of relationship (or association) present in a pair of variables. Each variable, in turn, may contain dozens or hundreds of individual values. Or, as another illustration, a "factorial analysis" may be made on a series of experimental outcomes in a scientific or engineering study. From this, the relative effect and importance of a great number of variables affecting the outcome of the experiment can be stated explicitly. Not only do we have data parsimony as a result, but we also gain insights that could not be obtained by mere unsophisticated view of the detailed experimental data by themselves. It is not surprising, therefore, that such analysis is also used widely in such business areas as production management and market research. But for every application that is made, there are several that should be performed but are not—simply because management has not had the time, opportunity and recognition of the importance of the approach. Thus, to the degree that there is developing a greater and deeper managerial awareness of the powers, values, and benefits of such analysis, to that extent will there be further growth in the usage of these truly time-saving and insight-furthering techniques.

In the light of the 1980s, looking back on the seventies, it may well appear that we have been indulging in some wasteful practices. We have not used to the fullest extent the known techniques for economical yet powerful data analysis and presentation. We probably have not pushed research sufficiently. And our computer software and hardware may be found to have been all too rudimentary. In the meantime, however, there is a wide range of uses—from relatively sophisticated to none at all—of good applications of charts, graphs, and tables as a thought- and information-conveying medium. To the extent that the present volume serves the efficient and creative use of visual and tabular materials, to that degree will it have been of service.

Glossary

It is always useful to have a quick reference in which the principal terms developed in a book are compiled for ready review. It must be recognized, though, that the definitions given cannot be detailed, nor can they be supplemented with illustrations.

A perusal of this glossary will reveal that a good many terms have been included that are new, in the sense that they have not occurred previously. Such was done advisedly. In the body of this book, we have striven for conciseness and clarity, sticking to essentials, and avoiding the clutter of unnecessary excess of technical terms. For those readers who wish to continue further readings in the area of chart and graph preparation, a brief note and definition of these additional terms will be helpful, for which the glossary seems the logical place.

Terms and definitions are generally interdependent. Often there are synonymous or overlapping meanings, and the definition of one concept involves reference to other terms. A glossary recognizes this relationship. In the pages that follow, therefore, words that are in italics indicate technical terms that are defined elsewhere in the glossary.

As in any technical field, terms and concepts are often borrowed from general language to convey a specific meaning. The glossary does not intend to bring the total definition of each term, merely that definition which is germane to graph and chart terminology.

Abscissa   The horizontal axis or base line of a chart.
Accented grid lines   Lines in a grid that have been emphasized by making them heavier. Often every fifth or tenth line is accented.
Activity chart   A type of *simo chart,* with emphasis on combination of hand-machine or multioperator processes.
Alignment chart   Same as *nomogram.*

Array   A regular arrangement of statistical data, often in sequence from largest to smallest, or vice versa.

Arrow diagram   Chart showing the sequence of interlocking operations in a physical system, from start to finish of the project. Also known as a *project graph.*

Axonometric projection   Projection of a three-dimensional figure in such a way that all lines of the object depicted are drawn to exact scale. Some distortion results.

Bar chart   A chart utilizing a series of bars to depict data. Bars may be arranged horizontally or vertically.

Binary decision flow chart   A *decision flow chart* in which the possible actions taken are of an "either/or" type, permitting just one of two choices.

Block diagram   Schematic of the layout of areas or units in block form.

Breakeven chart   Graphic analysis of the behavior of costs and revenue functions, in order to determine that level of operations at which total revenue equals total cost.

Cabinet projection   Distortion-free axonometric projection by reducing oblique angles to roughly half scale. See *axonometric projection.*

Calculation chart   Schematic of calculation procedures, usually accompanied by some pictorial elucidations, representing either the data analyzed or the calculation procedures involved.

Chart   Sheet exhibiting data concepts of processes in graphic form. The term may refer to any methodical arrangement, thus including tables and maps. It is often used as an alternative to *graph* or *diagram.*

Circular percentage chart   Same as *pie chart.*

Class   A group of values. For example, ages of men may be grouped into the classes 0 to 9, 10 to 19, 20 to 29, etc.

Class interval   The range of a *class.* For example, when a class ranges from 20 to 29, the class interval is $1 + 29 - 20$, or in other words, 10.

Class mark   The midpoint of a *class interval.*

Coffeegrinder   Slang term, applied to flow charts of procedures, particularly those forms that are vertical in design, with inputs in upper section and outputs in lower section, with appropriate feedbacks from outputs back to inputs.

Column diagram   Same as *histogram.*

Conceptual visualization   Diagram depicting the relation among various concepts that together form a total system, philosophy, idea, or other complex of thought.

Continuous curve   An unbroken curve or line.

Control chart   Time series graph containing an upper and/or lower control limit. When a plotted point falls beyond these limits, this fact signals the need for a study of causes responsible for the "out-of-control" condition.

Cybernetics   See *feedback.*

Cross-section   A network of horizontal and vertical lines. When the lines are equidistant and cut across each other at right angles, they form a *quadrille ruling*.

Data scale   Graduated reference line at the base and sides of a graph, marking off a progression of values at properly spaced distances. The markings are called *ticks*.

Decision flow chart   Pictorial representation of a decision process. When only yes/no types of decisions are involved, the chart represents a *binary decision flow chart*.

Dendrite   Branching figure or marking, such as a tree diagram.

Dependency chain   A sequence of interdependent events, with interlocking requirements. As a result, some activities cannot occur until one or more prior tasks have been satisfactorily completed. A dependency chain grapb depicts these constrained relations.

Diagram   A drawing that illustrates the relationship among a set of data, outlines a flow, describes a process, depicts an operation or procedure, or sets forth a plan. The term is often used in place of *chart* or *graph*.

Discrete curve   A curve or line with breaks.

Dot diagram   A visual presentation of the frequency of occurrence of individual measurements in a set. Each occurrence is denoted by a dot.

Fanfold   An oversize diagram that folds out. Also, a pad of several transparent diagrams mounted on top of each other, revealing separate layers of an illustration.

Feedback   The returning of part of the information output from a system of information or procedures flow, as new data serving to appropriately modify, adjust, or correct the input. Feedback thus serves as a cybernetic or control device. When preparing flow charts, feedback is often an essential aspect of proper representation of information or procedures flows.

Flow chart   Schematic of materials processes, operations, procedures, or information through a series of stages.

French curve   Flat drafting template consisting of scroll-like curves.

Frequency diagram   A diagram representing the frequency of occurrence of individual measurements in a set of measurements. Examples of frequency diagrams are the *histograms* and the *frequency polygon*. When the data forming a frequency diagram are cumulated, we obtain a cumulative graph known as an *ogive*.

Frequency distribution   A compilation of data, denoting the frequency of occurrence of values in a set of classes, and depicted in the form of a frequency diagram. Frequency distributions may be continuous or discrete and they may represent actual data or conceptual values. A principal example of the latter is a distribution of probabilities of occurrence of certain events.

Frequency polygon   A visual presentation of the frequency of occurrence of individual measurements in a set. The points on the graph, representing each frequency, are joined by a line. The word "polygon" refers to the many angles or corners formed by the joining of the points which are generally at various heights above the base line.

Gantt chart   Diagram for scheduling operations and processes. Projects scheduled for completion are identified in the left-hand section of the chart, with the remainder sectioned off in accordance with a linear time scale. On this scheme, "jobs" are then loaded in terms of anticipated starting and completion dates. Unloaded segments of the chart indicate available time for "scheduling-in" additional projects.

Gimcrack   A pretty-looking but empty chart, where "empty" means "devoid of real content."

Gozinto chart   Formed from the words "goes into," the diagram is a type of *flow chart,* showing how individual parts and components go into subassemblies and ultimately into the total product.

Graph   A series of points, either discrete or continuous, forming a line, curve, or other representation of data and information.

Grid   A network of lines, forming the background on which a graph is drawn.

Grouped bar chart   A bar chart using groups of bars for each *class* or other interval on the reference scale. The bars are distinguished by different colorings, shadings, or cross-hatching.

Histogram   A visual presentation of the frequency of occurrence of individual measurements in a set. Each frequency is shown by a bar, with all bars joined sideways.

Illustration   Device for conveying information pictorially, by drawing, printing, or projection.

Index chart   A line chart on which a time series in index form is plotted. For example, assume that an industry's labor force consisted of 2 million people in 1960, 3 million in 1970 and is expected to be 4 million in 1980. If we call 1970 the base year and set it equal to an index of 100, then the index figure for 1960 is $100 \times 2/3 = 66.7$. For 1980, the index will become $100 \times 4/3 = 133.3$.

Isometric ruling   Graph paper grid facilitating the drawing of horizontal edges of an object at 30-degree angles, while verticals are drawn perpendicular to the base.

Key   A legend, associated with a map, chart, or graph explaining the meaning of symbols used.

Layout   Chart showing the actual or proposed arrangement of components, equipment, or other elements of a physical or conceptual system.

Least-squares line   A *regression line.* The term "least squares" is based on a particular characteristic: the vertical squared distance of each plotted point from the line is at a minimum.

Legend   Inscription on a map, chart, or graph. Usually, the legend appears under a graph, giving title and explanatory text regarding the material presented. It may also be a key.

Line chart   A graph on which data are shown by a line, usually connecting a series of points. A common form is for the horizontal axis to show a time scale, while the vertical axis shows values in dollars, pounds, temperature, or other units of measurement.

Logarithmic paper   Graph paper with a grid scaled in logarithmic increments. Such a form is particularly suited for plotting *ratio-charts*.

Mark   A point-indication on a scale, dividing the length into main sections and subdivisions thereof.

Matrix   A rectangular arrangement, in row and column form, of a set of statistical or other types of data.

Mean   Average or center of the *frequency distribution*, also known as "arithmetic mean."

Median   Mid-point of a *frequency diagram*.

Mode   Most frequent value in a frequency diagram, hence its peak.

Multi-factor chart   A series of individual effects graphs are combined in a joint overall chart, with the purpose of showing the response of the dependent variable.

Multiple surface chart   Same as *stratum chart*.

Nomogram   A set of numerical scales in graphic configuration, showing the relation among the scalar variables.

Nomograph   Same as *nomogram*.

Oblique coordinates   A system in which the coordinate axes meet at other than right angles.

Oblique projection   Projection in which that face of a three-dimensional object which is parallel to the plane is drawn to exact proportions, but the remaining faces are not drawn at right angles.

Ogive   A cumulative *frequency diagram*.

Operation sheet   Chart showing in word and diagram a sequence of operations.

Ordinate   The vertical line of the chart, perpendicular to the base line and containing the vertical scaling.

Organization chart   Chart outlining the essential relations of a structure, such as the organization of a firm as represented by the principal titles and functions of the executives, managers, supervisory, and staff groups.

PERT network   Chart, similar to an *arrow diagram*, showing the interlocking nature of a set of activities forming part of a major project. Time values for completion of each activity are shown and a "critical path" of time requirements is traced through the network.

Pictograph   A type of bar chart, with each column consisting of a line-up of pictorial representation of such units as people, cars, houses, or other items of interest. Each pictured unit represents a stated magnitude, such as 100, 10,000, 1,000,000 or other value.

Pie chart   A circle representing a whole which is broken down into its various component parts. A pie chart may show absolute values, such as dollars, pounds, or other units or it may show relative values in percent. In the latter case, since the whole must equal one hundred, the chart is a "one hundred percent pie chart."

Plate   An illustration of either full size, over-sized, and folded form or a distinctly marked and bordered subsection of a graph. For example, if a graph is segmented or broken up into four subsections, then each of the subsections may be termed a "plate" of the graph.

Polar coordinate paper   Special type of graph paper, circular in form, on which points are plotted in terms of a vector and the angle that this vector forms with a reference line.

Probability distribution curve   A diagram representing the frequencies of a probability distribution, such as the normal distribution and resultant curve, or the binomial distribution and resultant histogram. A frequency diagram changes from polygon or histogram form to a smooth curve when the intervals between points become infinitely small.

Probability paper   Graph paper in which the grids are scaled in proportion to increments of some probability distribution (see also *probability distribution curve*.) Popular probability graphs are the Poisson, normal, binomial, and Weibull charts.

Procedural flow chart   Flow chart depicting the procedures associated with the operation of a system.

Process flow chart   *Flow chart* depicting a production process.

Project graph   See *arrow diagram*.

Quadrille ruling   A gridwork of equidistant horizontal and vertical lines, cutting each other at right angles, thereby forming an effect of equal-sized squares throughout the background of the graph.

Ratio-chart   A chart that utilizes rulings that are spaced in direct conformity with a set of proportional relationships. Usually, a grid with *semi-logarithmic scaling* will result in this proportional effect.

Rectangular coordinates   Two straight reference lines, perpendicular to each other, forming the *ordinate* and *abscissa*.

Regression line   A least square line or curve showing the average relationship among a set of points forming a *scatter diagram*.

Scale caption   Identification of the units of measurement, time performance or other values, as well as the general title applied to a scale.

Scaling indication   tick.

Scatter diagram   A plot of points reflecting the relation between two variables. Each point shows, for a given value of an independent variable (usually scaled along the abscissa) the corresponding value of the dependent variable (scaled along the ordinate).

Scatter plot   Same as *scatter diagram*.

Schematic   A representation, in diagram form, of a plan, method of operation of flow process.

Silhouette chart   Line graph in which the areas of deviation from a base line are shown in shaded form, giving a silhouette effect.

Semi-logarithmic paper   Graph paper in which only the vertical or horizontal scaling, but not both, are scaled in logarithmic increments. See also *logarithmic paper*.

Simo chart   A *flow chart*, emphasizing the coordination of two simultaneous (hence the word "simo") manual operations.

Slide   Transparency for projecting charted information.

Step-series graph   One of a series of graphs showing in successive steps how a number of ideas or concepts leads to an overall finding in the final graph.

Strata chart   Plural form of stratum chart.

Stratum chart   A graph that reveals the components of a *line chart*. The components are generally shown as bands with distinct shading, coloring, cross-hatching, or other separation.

Stub-mark   Same as tick, a very short projecting heavy line, at periodic points of a scale, serving to indicate the scale mark.

System operation chart   Flow chart depicting the flow of work, operation materials or information in a system.

Tally   A check mark, noting the occurrence of a value within a group or *class*. When a set of data has been fully tallied, the results form a *frequency distribution*.

Template   Graphic aid, usually in transparent form, serving as a guide for *layouts* or other diagrams.

Tick   Markings on a *data scale*.

Tick-mark   Same as tick.

Time-dependent flow chart   A *dependency chain* with time as an important variable.

Time series   Data classified on the basis of time intervals. "Period data" represent values accumulated during a period such as a week. "Point data" pertain to a specified point in time such as a date.

Underlay   A drawing guide that is laid under the (transparent) paper on which the art work is done.

Vellum   Specially textured paper or cloth for drafting purposes.

Word chart   A chart containing words only, designed to emphasize key sentences or ideas.

Bibliography

AMERICAN SOCIETY OF MECHANICAL ENGINEERS, *Operation and Flow Process charts,* ASME Standard 101, American Society of Mechanical Engineers, New York, N. Y. May 1947, 13 pp.

Developed by an ASME special committee, the booklet recommends uniform procedures for charting of operator methods, materials flows, processing sequences, and their interrelations, as used in time and motion studies and in industrial engineering work.

AMERICAN SOCIETY OF MECHANICAL ENGINEERS, *Illustrations for Publication and Projection,* American Standard, ASA Y 15.1-1959, American Society of Mechanical Engineers, New York, N. Y. 1959, 14 pp.

This standard develops recommended practice for effectively designed charts and graphs. Standard sizes for illustration originals and for lettering are given, useful particularly when preparing lantern slides for projection.

AMERICAN SOCIETY OF MECHANICAL ENGINEERS, *Time Series Charts,* American Standard, ASA Y 15.2-1960, American Society of Mechanical Engineers, New York, N. Y. 1960, 83 pp.

A collection of "preferred practices," setting forth what is considered "best usage" in the layout, design, scaling, curve-drawing and labeling, and captioning of time series graphs.

BACHI, ROBERTO. *Graphic Rational Patterns, A New Approach to Graphical Presentation of Statistics.* New York: Israel Universities Press, 1968, 243 pp.

A new, flexible system for graphing statistical data, in terms of various combinations, sequences, and arrangement of graphic patterns, is presented. The system permits extensions to highly refined and comprehensive comparative presentations, particularly where dynamic processes are involved.

BOWMAN, WILLIAM J. *Graphic Communication.* New York: John Wiley & Sons, Inc., 1966, 208 pp.

A wide diversity of solutions are offered for recurrent problems in graphic communication, emphasizing (1) the graphic figure as a communicative

180

vehicle, (2) visual language from a viewpoint of spatial organization, image composition, and expressive functions, (3) principles of visualization, and (4) a design library as a resource of graphic concepts.

DEPARTMENT OF THE ARMY. *Standards for Statistical Presentation.* Pamphlet No. 325-10 (D101.23 325-10), 1966, 141 pp.

A comprehensive review of various forms of data presentation, together with descriptive titles and numerous illustrative examples.

INTERNATIONAL BUSINESS MACHINES CORPORATION, *Flowcharting Techniques* (IBM Data Processing Techniques Document No. C20-8152-0), undated. Data Processing Division, IBM, White Plains, N.Y. 34 pp.

Identifies programming and systems flowchart symbols, gives practical procedures, examples, and worksheets.

KING, JAMES R. *Graphical Data Analysis with Probability Papers.* TEAM Technical and Engineering Aids for Management, Tamworth, N.H., 1966.

Covers the major types of graph paper for use in analyzing, plotting, and evaluating probability, providing detailed guides and instructions.

————. *Probability Charts for Decision Making.* New York: Industrial Press, 1971.

An authoritative work on the use of probability grids (such as normal, Poisson, binomial, and Weibull distributions) for specialized analysis of engineering, economic, and business data.

LYDON, PAULETTE LE CORRE. *The Graphic Primer.* Pol Lydon Inc., Brooklyn Heights, N.Y., 1961, 256 pp.

Presents methods of chart preparation, with a large variety of illustrative examples of time series, frequency distribution, and pie charts. Right and wrong methods are discussed. The types of data involved are predominantly of a descriptive statistical nature.

MC DANIEL, HERMAN. *An Introduction to Decision Logic Tables.* New York: John Wiley & Sons, Inc., 1968, 96 pp.

A presentation of the methods and procedures involved in the preparation of decision-structure tables (such as in Chapter 9 of this book). Included are a consideration of computer programming of decision tables and an appendix of problems and solutions.

U.S. DEPARTMENT OF COMMERCE, BUREAU OF THE CENSUS. *Bureau of the Census Manual of Tabular Presentation* (a special report of the Bureau of the Census). Washington, D.C.: Government Printing Office, 1949, 265 pp.

Subtitled "An outline of theory and practice in the presentation of statistical data in tables for publication," this book provides guidelines for the arrangement and composition of tabular material. Included in the discussion are formal definitions of structural parts of a table, titling, spacing, numbering headnotes and footnotes, and ruling of tabular material.

WELD, W. E. *How to Chart Facts from Figures with Graphs.* Norwood, Mass.: Codex Book Co., 1959, 218 pp.

Emphasizes the graphing of business and economic data, with special consideration of stock market series.

ZELAZNY, GENE. *Designing Flow Diagrams.* New York: McKinsey & Co., 1969, 60 pp.

A detailed analysis, from conception to completion of the development and presentation of a flow diagram. Includes a consideration of data acquisition and related information gathering aspects.

ZELAZNY, GENE. *Improving Chart Design for Oral Presentation.* New York: McKinsey & Co., 1965, 35 pp.

A useful guide that gives basic principles.

Index

Artwork template   146
Aspect ratio   149

Bar chart   26, 42
  horizontal   27
  modified   42
  three-dimensional   44
Bicking, Charles A.   166
Binaries   86
Borders   20

Calculation chart   50
Check list   22, 127
Column diagram   27
Completeness of charts   4
Computer plots   170
  programming   91
Control chart   38, 170
Convergence chart   76
Correlation   60
  coefficient   63
  multiple   61
  three-dimensional   61
Cumulative chart   31
Curve characteristics   45
Cybernetics   4

Data analysis   167
Decision charts   86
  binary   87

Decision tree   93
Decision tables   137
Dependency chain   79
Design as a thought process   7
Design principles   8
Design skills   6
Dual scaling   57

Experiment factors   47, 49

Feedback   4
Flow chart   67
  procedural   71
  process   67
  proportioned   76
Forms design   161
Frequency polygon   28

Grid   16, 97
  isometric   98
  logarithmic   98
  typewriter   97

Histogram   29
Human relations   21

Index numbers   37
Isometric form   61

Layout forms   99

Lettering   9
Letter sizes   147
Lewis, Mrs. D. L.   152
Line chart   25
Logarithmic scale   57

Margins   10, 132
Mechanics   9
Multi-factor chart   47

One hundred percent chart   26
Organization chart   84
Outline map   98

PERT network   64
Pie chart   28, 44
Probability distributions   28
Programmed decisions   138
Projection   146
Purpose of charts   1

Regression line   60

Scales   12
Scheduling chart   64

Segmentation   73
Silhouette chart   42
Slides   149
Source credits   152
Special scaling   148
Step-series graph   52
Stratum chart   26, 42
Stylized form   76
Symbols   99
Synthesis chart   63
Systems operation chart   67

Tabular flow chart   127
Tabular material   117
   development   117
   features   119
   headings   120
   variations in usage   122
Television screens   150
Templates   147
Time series   31
Titling   15
Transparencies   150
Tree diagram   93

Viewing factors   149